Understanding Performance Flight Testing

Kitplanes and Production Aircraft

Hubert C. "Skip" Smith

Second Edition

McGraw-Hill

New York Chicago San Francisco
Lisbon London Madrid Mexico City
Milan New Delhi San Juan Seoul
Singapore Sydney Toronto

Library of Congress Cataloging-in-Publication Data

Smith, Hubert.
Understanding performance flight testing : kitplanes and production aircraft / Hubert C. "Skip" Smith — 2nd ed.
p. cm.
New ed. of: Performance flight testing. c 1982.
Includes bibliographical references and index.
ISBN 0-07-137679-8
1. Airplanes—Flight testing. I. Smith, Hubert. Performance flight testing. II. Title.

TL671.7.S5324 2001
629.134'53—dc21

2001037025

McGraw-Hill

*A Division of The **McGraw-Hill** Companies*

1 2 3 4 5 6 7 8 9 0 DOC/DOC 0 7 6 5 4 3 2 1

P/N 0-07-137680-1
PART OF
ISBN 0-07-137679-8

The sponsoring editor for this book was Shelley Ingram Carr, the editing supervisor was Daina Penikas, and the production supervisor was Sherri Souffrance. It was set in Slimbach per the PFS2 design by Kim Sheran of McGraw-Hill Professional's composition unit, in Hightstown, NJ.

Printed and bound by R. R. Donnelley & Sons Company.

To my mother and father, who made my career possible,
and to the little town of Dillsburg, where I grew up

Contents

Acknowledgments ix

Introduction xi

1 Effects of Atmosphere 1

2 General Flight Test Procedures 11
Methods of Obtaining Data 12
Gross Weight Determination 18
Weather for Flight Testing 21
Use of Equations 25
Presentation of Data 27

3 Airspeed Calibration 33
Airspeed Errors 33
Test Procedure 36
GPS Method 43
True Airspeed 46
Airspeed Calibration Example (Speed-Course Method) 48
Airspeed Calibration Example (GPS Method) 49

4 Stall Speed 51
Bank Effects on Stall 54
Test Procedure 55
Stall Speed Test Example 59

5 Takeoff Performance 61
Forces Acting 62
Test Procedure 63
Photo Recording Method 65
Velocity Calculation 68

Wind Correction 70
Altitude Correction 72
Takeoff Distance to Clear Obstructions 75
Takeoff Test Example 78

6 Climb Performance 83
Drag 83
Power Required 86
Power Available 87
Rate of Climb 90
Test Procedure 91
Rate-of-Climb Test Example 95
Time and Distance to Climb 96

7 Cruise Performance 101
Power Determination 101
Cruising Speed Test Procedure 105
Maximim Speed Test Procedure 107
Range 110
Range Test Procedures 113
Endurance 119
Cruise Performance Example 120

8 Descent and Landing Performance 121
Determining Glide Ratio 124
Landing 127
Landing Test Procedure 131
Glide Distance Example 135
Landing Distance Example 136

9 Computer Methods 139
Airspeed Calibration 141
Stall Speed 143
Takeoff Performance 145
Rate of Climb 152
Cruise Performance 157
Glide Performance 161
Landing Distance 168
Conclusion 171

10 Evaluating Modification Effects 173

Acknowledgments

I am very grateful to Capt. William L. Posnett, USN (Ret.), former Navy test pilot, for developing a number of the data reduction methods used in Chapter 9, and for reviewing that chapter. I would like to thank Micro Aerodynamics, Inc. and Sporty's Pilot Shop for providing pictures, as well as Cessna Aircraft Corp. and The New Piper Aircraft, Inc. for allowing me to publish performance information from their handbooks. I am indebted to Kristie Kalvin for her excellent typing services on this edition. Also, not to be overlooked is Karen Rider, who had the unenviable task of translating my handwritten notes into a typed manuscript for the first edition of this book back in 1982.

Introduction

In the late 1930s, Clarke Gable starred in the movie *Test Pilot.* He was dashing, daring, adventurous, rugged, and not too terribly bright. In appropriate garb he could equally well have played a cowboy, private eye, explorer, sea captain, or any one of dozens of heroes typical of that era. Of course, the helmet and goggles, the leather jacket and boots, and the white scarf unmistakably identified him as a pilot. In this role he threw airplanes around the sky, put them into screaming dives, and otherwise pushed them to and beyond their design limits, often to the chagrin of his employers. Sometimes he barely escaped by parachute as the airplane literally fell apart around him. Thus was established the Hollywood image of test flying.

Real-life testing (like most subjects exploited by the film industry) is quite different, however. New airplane designs are tested very carefully and methodically. They are flown only by the most skillful and experienced of pilots whose brains usually exceed their brawn. Their judgment must be honed to an even finer edge than their skill, and procedures must be followed to the letter.

In the early stages of flight testing, the airplane is very gently maneuvered to determine its degree of stability and control. Only after its handling qualities have been determined as acceptable is it pushed to test its endurance near design limits. If any problems arise, it goes back to the drawing board (or computer display, nowadays).

Once the airplane is proved to perform safely, it then progresses to another stage of flight testing. In this stage the airplane is carefully tested to see if it meets the performance that was expected of it when it was designed. This procedure is, therefore, appropriately termed, "performance flight testing." Aircraft manufacturers employ rather sophisticated and complex equipment to make this determination, particularly in the case of military or other high-performance aircraft.

For airplanes on the other end of the performance scale, there are many old techniques that are much simpler and work almost as well. Many of these can be employed by the average pilot with little technical experience beyond that normally required to obtain flight ratings. These procedures are pretty much devoid of thrills, involving maneuvers less hair-raising than those typically required to satisfy a commercial pilot examiner. They do, on the other hand, require some precision flying if fairly accurate results are to be obtained. Such precision would need to be at about the level expected in normal instrument flying.

But why go through a lengthy procedure to flight test an airplane that has already been tested by the factory? There are several answers to this obvious question. First of all, the manufacturer tests a brand new, tightly fitted and carefully rigged airplane. Aircraft deteriorate, aerodynamically, with age. Scratches and dents gradually appear; door, window, and other seals begin to leak; patches are added to repair minor damage; and the continual disassembly and assembly required for inspections cause minor distortion and misalignment of airframe components. None of these flaws means very much in itself, but collectively they can add up to a significant decrease in performance. There is also decreased power output with engine wear. New airplanes also have brand new engines when performance is measured. All of these factors can mean a noticeable difference between the performance of an airplane right off the production line and one 25 years old with 5000 hours on it.

Then, too, not all airplanes are exactly alike, even when new. Surely you have seen ads for used airplanes where the craft was extolled as "exceptionally fast for a Skybird II." Some of these claims may not be exaggerations. Certain airplanes just turn out to be slightly better performers than their more normal siblings. By the same token, some also turn out to be a bit more doggy. You never see *this* fact mentioned in ads, however.

So far, we have been discussing production airplanes. If you are completing a homebuilt, then the necessity for flight testing should be obvious. Even those built to the exact specifications of the designer will vary considerably with construction technique. And what homebuilder has not substituted here and there for costly or unavailable materials or parts? In fact, one of the big attractions of homebuilding is the freedom to customize the airplane. You usually have a choice of engines, propellers, wheels, instruments, and other accessories. The temptation to make further modifications is great. Changing from conventional to tricycle landing gear or adding wheel fairings can often be accomplished without completely redoing the original design. But these modifications cause changes in performance. Even the addition of a canopy to an open cockpit can make a difference. Consequently, very few homebuilts of the same basic design are really alike. If you fly one of these, you really should test it to determine its particular performance. Of course, this goes without saying if you are building an original design.

There is one more reason to go through a series of performance flight tests, and that is simply to learn more about the airplane's performance. Certainly, you can refer to the charts in the operator's handbook if you fly a fairly average production airplane. But the numbers here take on a deeper meaning when you have run through the tests and verified them for yourself. There are also a number of things revealed that do not show up in the standard performance charts. Few pilots know such things as the rate-of-climb when flying at other than best rate-of-climb speed or the L/D ratio at various airspeeds. Even the maximum L/D ratio is an obscure number to veteran throttle pushers. The variation in stall speed with gross weight or the change in takeoff distance with density altitude will become very apparent as you plot it out yourself. In short, testing your airplane will enable you to understand its performance a lot better and will, consequently, make you a safer pilot.

This book describes a series of tests by which most of the performance information normally required for airplane operation can be obtained. The tests are designed to be carried out by the average light airplane pilot or someone with equal aviation background. No special equipment is required other than a stopwatch and a basic electronic calculator. In some cases, such items as a tape recorder or camera are helpful but not absolutely essential. The standard instruments in the airplane are the primary devices used to obtain the necessary measurements. For many of the tests only an altimeter, an airspeed indictor, and an outside air temperature gauge are required.

The first two chapters deal with background material and general procedures to be followed in conducting the tests and obtaining and reducing flight data. The next chapter explains the calibration of the airspeed indicator so that accurate information on airspeed is available for the subsequent tests. Chapter 4 outlines tests for the determination of stall speed, which is also needed before some of the other tests can be performed. The next four chapters deal with the primary performance areas of takeoff, climb, cruise, and descent and landing, respectively. In addition to describing the flight test procedure for each of these areas, some discussion is included to explain the reasoning behind the procedure and why the airplane performs as it does. This information is presented in a very basic way and does not involve complex mathematics or require any special technical background for understanding of the material.

A more streamlined approach to data reduction and presentation is given in Chapter 9 for those who prefer to work with digital computers. All of the performance items in the previous chapters are covered here. The data are assumed to be collected in the same manner as before, but, instead of performing numerous hand calculations, they are simply entered into spreadsheet programs. The computer then converts the raw data into useful performance information, and plots it out in easily readable charts. Microsoft Excel is utilized for these programs. In the back of this book, a diskette containing the software for all of these programs, including sample spreadsheets and charts for each of the tests, is provided. The final chapter deals with evaluating performance before and after aerodynamic modifications. Quick and easy methods are presented here for those who do not wish to go through extensive tests for complete performance measurement.

An appendix is included which lists some useful information, and also some actual performance data on production airplanes, as presented in their operating handbooks. Upon completion of the tests outlined, you should be able to construct a similar set of charts custom tailored to your own individual airplane.

Warning

Users of data as obtained in these tests are cautioned to interpret it with a bit of wisdom. The book is intended just as the title suggests: *understanding* performance flight testing. You can use the performance data as a *guide* to the operation of your airplane, but don't expect the performance to be exactly as you predict. It would be rather foolish, for example, to attempt a takeoff on an 1800-ft runway simply because your tests show that 1790 ft are required under similar conditions. While all of the tests are based on sound scientific principles, many approximations are used. Also, remember that performance data in published pilot operating handbooks are based on extensive testing by experienced test pilots and flight test engineers. Numerous tests are run at each data point, and carefully analyzed to make sure that the results are correct. Even with the most careful testing that you might do, there is still some degree of error introduced. So, use some judgment in your application of the data, and allow adequate margins of safety. Neither the publisher nor the author will accept responsibility for the accuracy of the data as obtained from the tests described herein. Similarly, Piper, Cessna, and other contributors will not be liable for respective information submitted.

Fig. I-1. The performance of homebuilt aircraft like this Van's RV-6 is pretty much unknown until proven by flight tests (*Courtesy Ray Houtz, builder*).

Fig. I-2. Older aircraft such as this 1953 Bonanza are good candidates for performance reevaluation due to aerodynamic deterioration over the years.

CHAPTER 1

Effects of the Atmosphere

In addition to the characteristics of the airplane itself, performance depends a great deal on the nature of the atmosphere through which the airplane flies. Unfortunately, the atmosphere varies quite a bit in its makeup from day to day, and also, with altitude on any given day.

The atmosphere is a big mass of air which covers the earth like an ocean covers the ocean floor. Unlike the sea, however, the atmosphere gets less dense, or lighter in weight, as altitude is increased. This characteristic is due to the fact that air is a compressible fluid (water is not). The air is held against the earth by the attraction of gravity, which means that it has a certain amount of weight to it, and it is compressed by this weight. It is compressed most at the earth's surface because all of the atmosphere is above that level and all of it is bearing down on the layer right at the surface. As we go higher, say to 50,000 ft, only the atmosphere *above* 50,000 ft is bearing down on that level. The air up there is compressed less. The same volume of air has less weight at altitude than at the surface because of this lower compression.

The measure of how much weight a certain volume of a substance has is referred to as its *density.* A cubic foot of water weighs 62.4 lb; a gallon of gasoline weighs 6 lb. We call this value of 62.4 lb/ft^3 (pounds per cubic foot) or 6 lb/gal the density. These fluids, being incompressible, have a fixed density. They can vary *slightly* with temperature, but at normal temperatures are always very close to these weights. Air, on the other hand, is extremely variable. There is no fixed weight per cubic foot associated with it. Air density is very important, however, since all aerodynamic forces, including lift and drag, directly depend on it. The engine power output is also proportional to the air density in which it operates. We must, therefore, know the density of the air in order to determine the airplane's performance.

Density of the air cannot be measured directly. Fortunately, though, density depends on some other properties of the air, namely pressure and temperature, which can be measured. Density is directly proportional to the pressure of the air and inversely proportional to the temperature. This means that the higher the pressure, the higher the density, but the higher the temperature, the lower the density will be.

Mathematically, this relationship is given as:

$$\rho = \frac{P}{T} \times \text{constant}$$

This equation says that ρ, the Greek symbol used for density (pronounced "roh") is equal to pressure divided by temperature multiplied by a constant (a number). Thus once pressure and temperature are known, the density can be calculated.

Pressure and temperature, of course, vary from day to day, from one location to another, and with altitude. However, there are certain average values of these quantities that exist at the earth's surface. Also, since the altitude of the surface varies, there is a standard reference base established from which altitude is measured. This reference is the average height of the sea and referred to as "mean sea level" or MSL. The average conditions established as standard at mean sea level are:

$$\text{P} = \text{pressure} = 29.92 \text{ in Hg (inches of mercury)}$$

$$\text{T} = \text{temperature} = 59^\circ\text{F}$$

These two values will also give a standard density when used in the equation given above of:

$$\rho = \text{density} = 0.0765 \text{ lb/ft}^3$$

This would actually be the density at the surface of the earth. If we were on the moon (not anticipated in the following flight tests), the density would be different. A cubic foot of air would weigh less. A better measure of density is the *mass* density. Mass is weight divided by the gravitational constant, or more properly, weight is mass times the gravitational constant. Hence, mass is the same everywhere, but its weight depends on where it is with respect to some gravity acting upon it. The mass density is really what is used in performance calculations, and has units of slugs per cubic foot. A slug is actually $1 \text{ lb} \cdot \text{s}^2/\text{ft}$. Hence, the combined term, "slug."

In addition to these standard values at sea level, there is a standard rate of change of temperature with altitude. This rate is $-3.579°$F per thousand feet, and is referred to as the "standard lapse rate."

Using the standard value of temperature at MSL and the standard lapse rate, standard values of pressure and density can also be determined for various altitudes. On a *standard* day these values would really exist at those actual altitudes. Table 1-1 shows the standard values of temperature, pressure, and mass density at altitudes up to 20,000 ft. On a standard day, for example, at 6000 ft the temperature would be 37.6°F, the pressure 23.98 in Hg, and the density 0.001987 slug/ft^3.

On most days, however, the atmospheric properties are not standard. The pressure at 6000 ft MSL may not be 23.98, and the pressure at sea level may not be 29.92. However, there will be *some* altitude at which the pressure is 29.92 (standard sea level pressure), and 6000 ft above this, the pressure will be 23.98. This altitude is referred to as 6000-ft *pressure* altitude, regardless of the true altitude at which it is located. Pressure altitude is the height of a respective value of pressure above the height where 29.92 in Hg is located. Remember that only on a standard day this is the value right at sea level.

To illustrate pressure altitude, consider the airplane in Fig. 1-1. On a standard day, the pressure is 29.92 at sea level. If the pilot correctly sets this value into his altimeter it will read zero at sea level and 6000 ft at 6000-ft true altitude (assuming there is no instrument error). Now let us assume that the pilot flies into a lower pressure area. The pressure at sea level here is 28.86. If he resets his altimeter to this value he will continue to read 6000 ft at that altitude, and, if he were to descend to sea level, the

Table 1-1. Standard Atmospheric Properties

ALTITUDE FT	TEMP. °F	PRESSURE IN HG	DENSITY SLUGS/FT^3	DENSITY RATIO
0	59	29.92	0.002377	1
1,000	55.4	28.86	0.002308	0.971
2,000	51.9	27.82	0.002241	0.9427
3,000	48.3	26.82	0.002175	0.9151
4,000	44.7	25.84	0.002111	0.8881
5,000	41.2	24.89	0.002048	0.8616
6,000	37.6	23.98	0.001987	0.8359
7,000	34	23.09	0.001927	0.8106
8,000	30.5	22.22	0.001868	0.7859
9,000	26.9	21.38	0.001811	0.7618
10,000	23.3	20.58	0.001756	0.7384
11,000	19.8	19.79	0.001701	0.7154
12,000	16.2	19.03	0.001648	0.6931
13,000	12.6	18.29	0.001596	0.6712
14,000	9.1	17.58	0.001545	0.6499
15,000	5.5	16.89	0.001496	0.6294
16,000	1.9	16.21	0.001448	0.6092
17,000	−1.6	15.56	0.001401	0.5894
18,000	−5.2	14.95	0.001355	0.5701
19,000	−8.8	14.35	0.001311	0.5511
20,000	−12.3	13.76	0.001267	0.5329

altimeter would read zero. Notice now, though, that the standard sea level pressure of 29.92 doesn't exist at sea level but 1000 ft *below* sea level. Flying at 6000 ft in this area, the airplane is actually 7000 ft above the baseline altitude where the pressure is 29.92. Therefore, we say that the airplane, while at 6000-ft *true* altitude, is at 7000-ft *pressure altitude.* The pilot could easily determine this altitude by resetting the

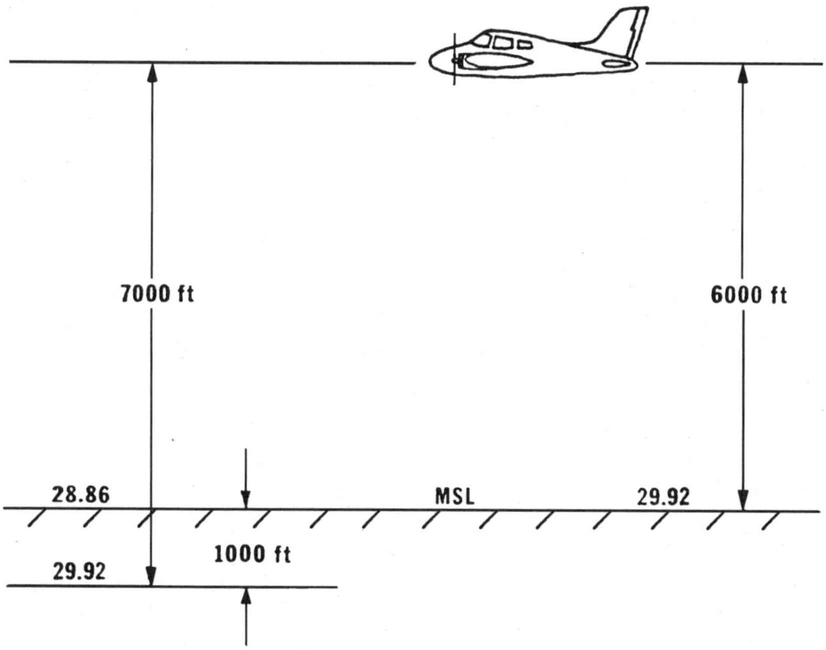

Fig. 1-1. Pressure altitude relationship showing the height of the airplane above the level where the pressure altitude is 29.92 in Hg.

altimeter to 29.92, in which case the altimeter would now read 7000 ft. Consequently, setting the altimeter to the actual sea level pressure gives an altitude reading in *actual,* or true, altitude; setting it to 29.92 gives a reading in *pressure* altitude.

This same sort of reasoning applies to another type of altitude, which is called *density altitude.* If the airplane were at 6000 ft on a standard day, it would be 6000 ft above sea level and also the level where the density is 0.002377 slug/ft^3. Again, however, let us consider a nonstandard situation. Suppose the combination of pressure and temperature at sea level resulted in a density of 0.002308 slug/ft^3. This is the density that occurs on a standard day at 1000 ft. In order to reach 0.002377, one would have to descend 1000 ft below sea level. Therefore, in this situation, at a true altitude of 6000 ft, the airplane is flying in density that normally occurs at 7000 ft as shown in Fig. 1-2. This altitude is thus referred to as 7000-ft *density altitude.*

Just as the actual density can be calculated from the pressure and temperature, density altitude can be calculated if the pressure altitude

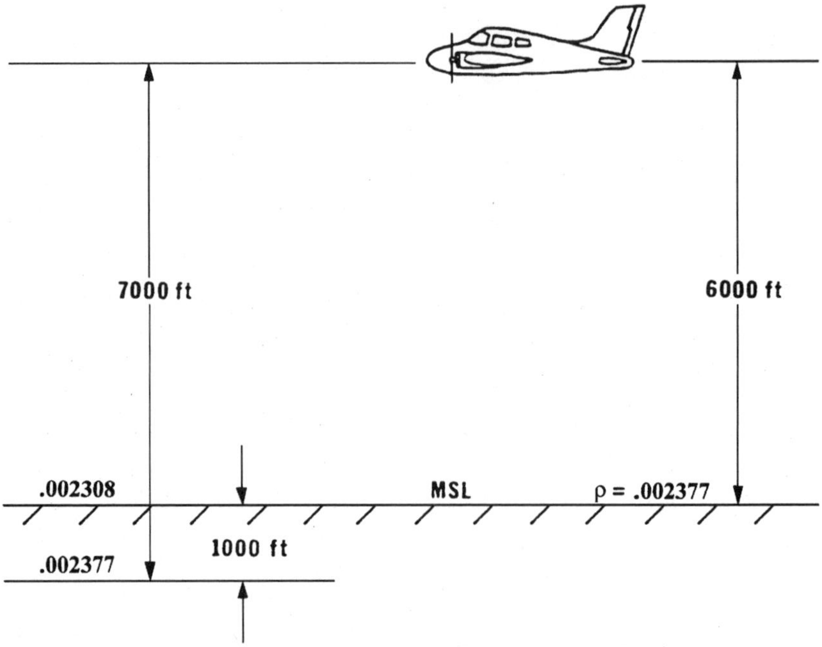

Fig. 1-2. Density altitude relationship showing the height of the airplane above the level where the density is 0.002377 slug/ft^3.

and temperature are known. We have already discussed the way in which pressure altitude can be determined from the altimeter. Finding the temperature is even easier. All that one needs to do is to read the outside air temperature gauge, or OAT, as it is called. The simplest way, then, to proceed to determine density altitude is to enter the appropriate temperature opposite the appropriate pressure altitude on the flight computer. The density altitude can then be read in the density altitude window of the computer (Fig. 1-5).

Determining density altitude in this way with an E6B type computer is rather inaccurate due to the very small graduations on the density altitude scale. A better way is to use the chart shown in Fig. 1-3. This chart is typical of many included in airplane information manuals. Pressure altitude and temperature must first be determined, as before. Then the chart is entered along the bottom with the appropriate temperature. The dotted line shows an example using 25°F. Go up vertically until reaching the line for the appropriate pressure altitude

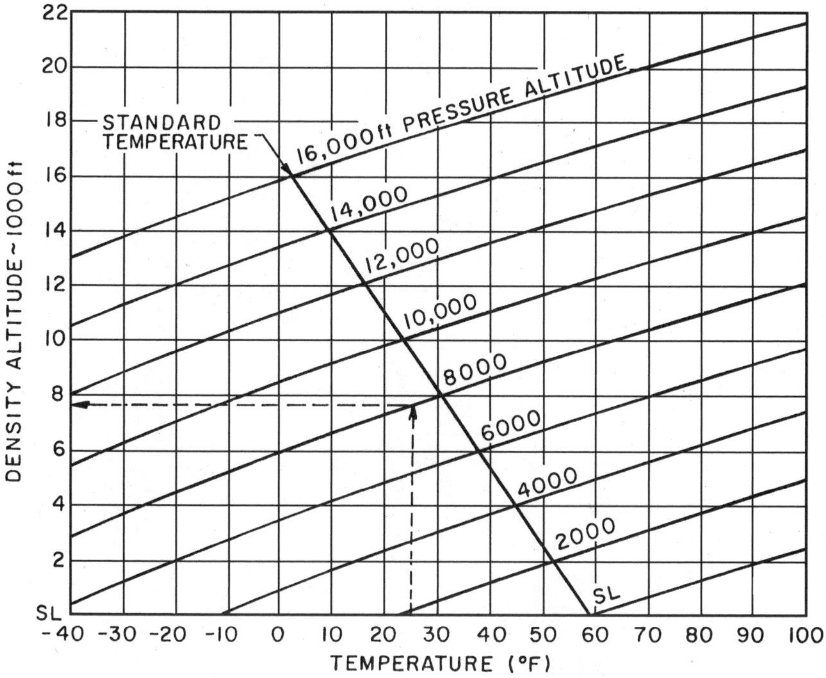

Fig. 1-3. Chart for determining density altitude from pressure, altitude, and temperature.

(in the example, 8000 ft). Then read directly to the left on the vertical scale, the density altitude. In the example, this is about 7500 ft. Be sure to use the proper temperature scale. The chart is plotted in degrees Fahrenheit, while most computers use degrees Celsius. A version of this same chart with finer grids for more accurate determination of density altitude can be found in App. A, Fig. A-1. An even better way of finding density altitude is with an electronic flight computer, which is further discussed in Chap. 2.

Occasionally, the actual density, or sometimes density ratio ρ/ρ_0, is required for a given altitude. The density ratio can be obtained from Fig. 1-4 once the density altitude is determined by one of the methods outlined above, such as the E6B shown in Fig. 1-5. The density ratio at 14,000 ft, for example, is 0.65. This means that the density here is 65 percent of what it is at standard sea level (or zero density altitude). The density, then, would be 65 percent of 0.002377 or 0.001545 slug/ft^3.

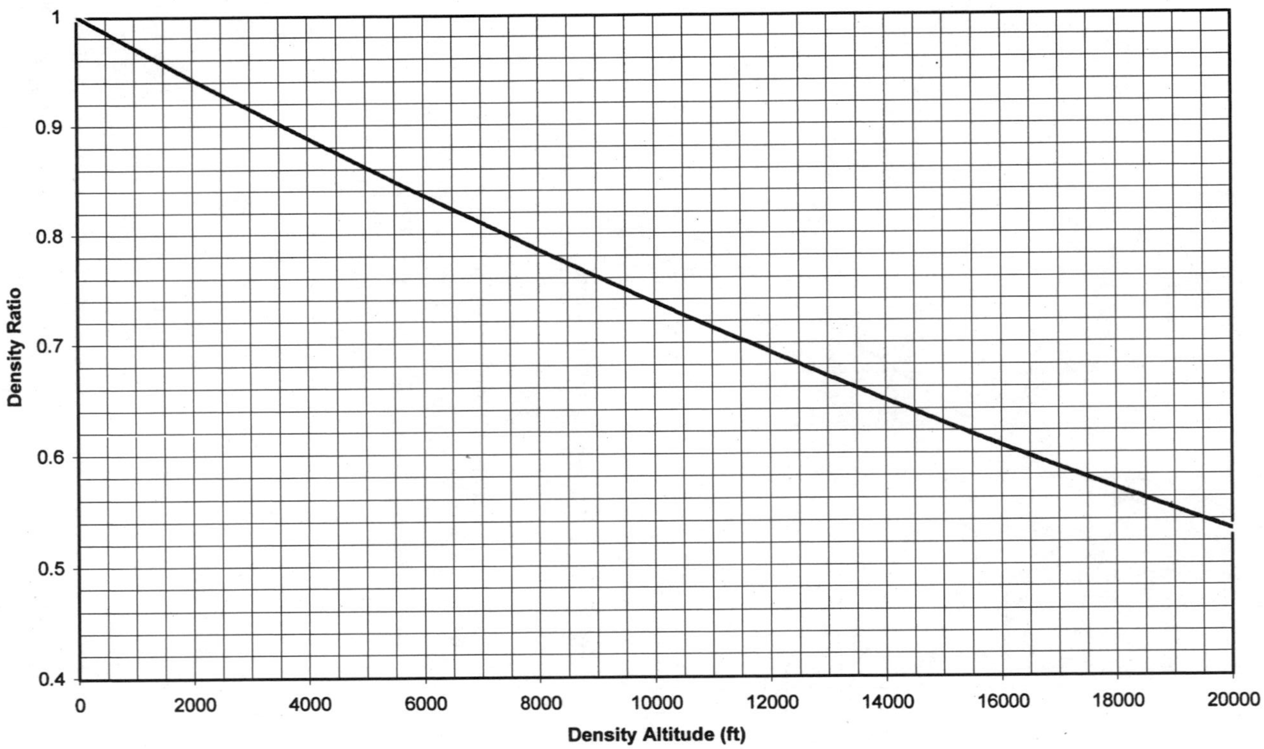

Fig. I-4. Chart for determining density ratio for the appropriate density altitude.

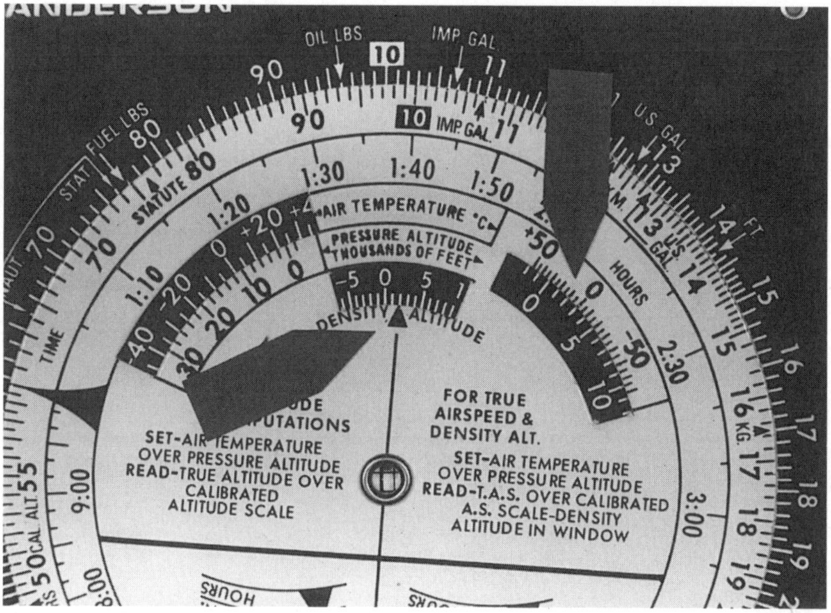

Fig. 1-5. Determining density altitude with E6B type flight computer. Setting temperature opposite pressure altitude yields density altitude in the density altitude window, as indicated by arrows.

While a little more complex math is involved, the actual equation to calculate the density ratio in the standard atmosphere is given in terms of the altitude h, in feet, as,

$$\frac{\rho}{\rho_0} = (1 - 0.00000689h)^{4.256}$$

This is really the equation of the curve plotted in Fig. 1-4. Since we don't usually operate in the standard atmosphere, for flight test purposes we often need to calculate the *actual* density ratio (rather than standard). This can be done if we know the pressure altitude h_p, in feet, and the outside air temperature T, in degrees Fahrenheit. This equation is even more involved, but gives a very accurate value:

$$\frac{\rho}{\rho_0} = \frac{519(1 - 0.00000689h_p)^{5.256}}{T + 460}$$

The density altitude h_d can also be calculated quite accurately from the density ratio as,

$$h_d = \frac{1 - (\rho/\rho_0)^{0.235}}{0.00000689}$$

These equations are employed extensively in the computer programs described in Chap. 9. If you are adept at working with equations, you may want to use them in hand calculations. If not, you can find density altitude from Fig. 1-3, or by use of an electronic flight computer, or even an E6B computer. Once density altitude is determined, the density ratio can be read from Fig. 1-4.

CHAPTER 2

General Flight Test Procedures

As with any flight operations, the flight tests described in this book must be conducted within the confines of the Federal Aviation Regulations and the operating limitations of the airplane. None of the tests require exceeding either of these restrictions. However, it is essential that the pilot observe all of the rules. The airspeed calibration test, for example, demands a low altitude, but FAR 91.79 regarding minimum safe altitudes should be followed. Likewise, a safe terrain clearance altitude should be attained before performing tests such as stall speed. It would be wise to observe FAR 91.71 dealing with aerobatic flight for all altitude flight tests, even though many do not necessarily involve abrupt altitude change maneuvers.

All test flying should be performed in good VFR weather; that is, not just marginally VFR. The tests should be conducted in non-congested airspace. Even so, it is particularly important to be alert for other traffic. The nature of test flying requires a pilot's concentration on the instrument panel in order to hold exact conditions, much like IFR flight. The observer(s) will also be concentrating on reading instruments, recording data, or spotting landmarks. It is a very good idea to also carry one more crew member than necessary for carrying out the tests, if space permits. This person can serve as a traffic spotter and should be someone familiar with this sort of activity.

If some crewmembers are not pilots or not accustomed to flying in light aircraft, they should be thoroughly briefed on safety and emergency procedures. Make sure that they understand how and when to use seatbelts. It is also wise to brief them on the meaning of warning horns, such as stall and gear warnings which may sound, or other unusual situations that may occur during the course of conducting some tests. This helps to relieve much anxiety in such situations and makes the testing easier and more pleasant. A calm data-taker also records much more accurate information than a nervous one.

Methods of Obtaining Data

More than one crewmember is required for most tests. Make sure that the duties of each are understood prior to takeoff (Fig. 2-1). The pilot needs to concentrate on accurately flying the airplane. A second crewmember may need to continually observe instrument indications. A

Fig. 2-1. Planning and coordinating duties among crewmembers prior to flight test.

third person is often desirable to record data read out by the observer. In other tests, landmark spotting is required by one person. Another may serve as both instrument observer and data recorder in such situations. Usually, three people are required to comfortably perform the tests. Most tests could possibly be performed, however, by two and would have to be in the case of a two-place airplane.

Data are obtained, in most cases, from the standard instruments installed in the panel. If the observer is not a pilot, make sure that he or she is familiar with the instrument indications. Also, make sure that everyone in the crew understands what data will be taken and in what order. It would be wise to rule off columns on a sheet of paper to be used as a data sheet. Label the column headings for the data to be taken in the order in which it will be read. A well-coordinated effort among pilot, observer, and recorder will make testing much easier and avoid confusion later when the data are reduced and plotted. Appendix A contains a typical data sheet (Figs. 2-2 and A-2).

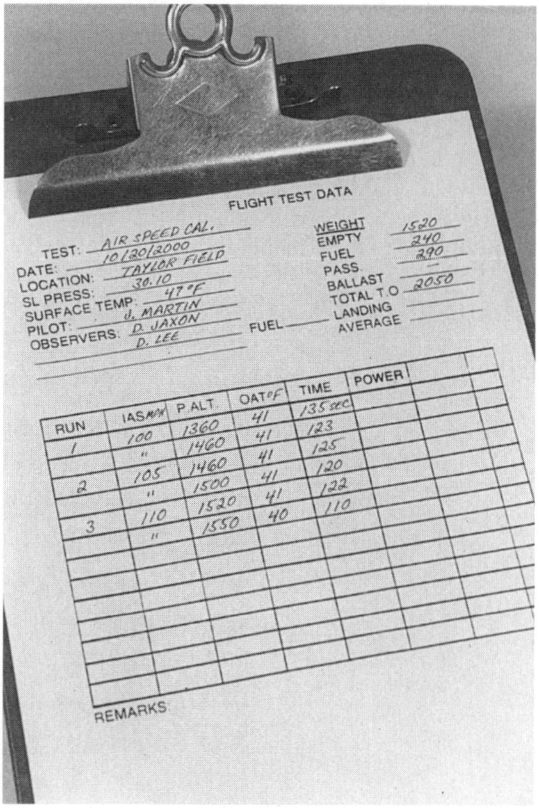

Fig. 2-2. Data should be recorded on a formal data sheet, as provided in App. A.

Practically all tests will require an airspeed value. Usually, the true airspeed is ultimately required. However, the value as read from the air speed indicator is the indicated airspeed (IAS). It is the value that you must start with in order to calculate a true airspeed. Methods for correcting to calibrated and true airspeed are discussed in Chap. 3. Some airspeed indicators are calibrated in knots and some in miles per hour (Fig. 2-3). Other indicators have both scales. It doesn't really matter which scale is used as long as you are consistent. If both scales are indicated, it is probably best to read the outer scale, usually in statute miles per hour, simply because the graduations are larger and, hence, the reading will be more accurate. If airspeed information is desired in knots, miles per hour readings could be converted later.

Fig. 2-3. Typical airspeed indicator. When two scales are given, the outer scale (mph here) should be used, since it is easier to read accurately.

Performance of the airplane is also highly dependent on the density of the atmosphere as discussed in Chap. 1. Since density can't be measured directly, we depend on measuring temperature and pressure in order to calculate density. Pressure is actually measured in terms of pressure altitude. The altimeter is used for this purpose and indicates in *pressure* altitude whenever the barometric pressure in the altimeter window is set to 29.92 in Hg. Since pressure altitude (not true altitude) is what is required in all fight tests, the altimeter should always be set at 29.92 during the testing (Fig. 2-4). Takeoff, landing, and other flight operations in getting to and from the test area, however, require a knowledge of *true* altitude. Therefore, it is best to set the altimeter to the proper barometric pressure at the time and location (or field elevation) prior to takeoff. After takeoff and climb to the test location, the altimeter can be reset to 29.92. It is, also, a good practice to record the true pressure so that the altimeter can be reset to true altitude for descent and landing. Read the altimeter and record the data in feet of pressure altitude. Since the smallest increment on most altimeters is 20 ft, the nearest 20 (or possibly 10 ft) of altitude is sufficient for accuracy. Ten feet or so of altitude makes very little difference in density.

Temperature is obtained from the outside air temperature gauge, or OAT. Most of these gauges also have two scales, one in degrees Celsius and one in degrees Fahrenheit (Fig. 2-5). It is again more convenient to read the larger scale, which is the Fahrenheit scale. The charts and calculations in this book are thus based on temperature reading in degrees Fahrenheit. Read the temperature as closely as possible to single degrees, for example, 54° or 57°, even though the smallest increments on the scale are 5°.

For fuel consumption determination in calculating range, the fuel flowmeter can be utilized if one is installed in the panel. Such meters

Fig. 2-4. Altimeter showing pressure altitude when set to 29.92 in Hg.

Fig. 2-5. Typical outside air temperature gauge. The outer scale (Fahrenheit in this case) is larger, and thus more easily read.

are usually standard on airplanes with fuel-injected engines. This gauge reads either in gallons or pounds of fuel per hour. Although not required to calculate the performance, it is necessary to know the proper power settings for a certain percentage of rated power in the cruise performance tests. Power is set by the appropriate revolutions per minute with a fixed-pitch propeller or combination of revolutions per minute and manifold pressure with a constant-speed propeller. Therefore, attention should be given to the tachometer and manifold pressure gauge to ascertain correct power settings for cruise speed and range determination (Fig. 2-6).

To summarize, the following instruments will be utilized to obtain the respective information:

- Airspeed indicator—indicated airspeed
- Altimeter—pressure altitude
- Outside air temperature gauge—temperature

Fig. 2-6. The tachometer and manifold pressure gauge, used for setting specific amount of engine power. Also included below manifold pressure is the fuel flowmeter, used to determine fuel consumption.

- Tachometer and MAP gauge—engine power
- Fuel flowmeter—fuel consumption rate

These instruments are shown in Figs. 2-3 through 2-6. For many tests, only the first three of these instruments will be required to yield the necessary information for measuring performance. In some tests some timing must be done and is best performed with a hand-held stopwatch. A portable tape recorder and a camera are also useful for recording data in certain tests. However, these devices are not essential. You may want to use your ingenuity and come up with some original devices or applications of devices for variations on some of the testing techniques.

In the event that you want to test a single-place airplane, the procedure is a bit different. Obviously, the pilot is the whole crew. He will have to fly the airplane, do the observing, and record the data. Needless to say, he will have his hands full. However, this feat is not impossible. If you are the pilot (and crew) of a single–place, it would be desirable to use a tape recorder for recording much, or all, of the data. This procedure eliminates the distraction from observation of the instruments while writing down the reading. A stopwatch mounted on the panel or yoke, at least temporarily, would also be a big help. Develop your own sequence of reading and recording information which seems to be most logical and convenient for you. Don't forget that you are also the traffic observer in this case. Pay particular attention to choosing a safe area and time for testing. It would be wise to also make frequent clearing turns between runs or steps in the test flights.

Gross Weight Determination

The weight of the airplane has an effect to some degree on all items of performance. The effect is quite significant on certain performance items such as takeoff, climb, and stall speed. For these tests, therefore, it is important to know the weight rather accurately. In other tests, such as maximum speed and range determination, the weight has a lesser effect and some variation in the weight can be tolerated.

Total weight, of course, means the entire weight of the airplane as flown, including fuel and payload at any particular time. Total weight is sometimes referred to as *gross* weight. There are maximum gross weight limitations on each airplane type. Sometimes these limits are different for takeoff and landing. For most single-engine airplanes, though, they are usually the same. In this case, the value is referred to simply as the maximum gross weight.

Some performance items, such as climb and cruise, are usually presented in operator's manuals for one gross weight. This weight is known as the *standard* weight, and for light single-engine airplanes is normally the maximum gross weight. Other performance items, such as takeoff and landing distance, are presented for a variety of weights. For the tests in this book, weight corrections can be made, in some cases, for weights other than standard. This is true for stall speed and takeoff and landing performance. For other performance items, it is necessary to establish a standard weight and make all tests as close to that weight as possible.

The question now arises as to what is standard weight. The maximum allowable gross weight could be chosen for this value. Another choice, and perhaps a more practical one, is a normal operating weight. Such "normal" weight would depend on your particular flying habits. If you *normally* carry four people and baggage in your four-place airplane, then the maximum gross weight is pretty normal. On the other hand, if typical flights usually involve only two people without baggage in a four-place craft, then the average weight in such configuration might be a more normal value for you. The advantage of using maximum gross weight is that this value would yield the minimum performance and any lower weight would improve upon it. In this way your performance is always on the conservative side. Rate of climb, for example, at lower weight would be greater than that at maximum gross and takeoff distances would be shorter. The choice is up to you, but then you must adhere to that weight in loading of the airplane during the course of the flight tests.

For all tests it is desirable to know the weight of the airplane fairly accurately. The best way to determine this is to fill the fuel tanks completely before each flight test. In this way there is no doubt about the exact amount of fuel aboard. Then record the weight of each crewmember. If any of the crew are uncertain of their weight, have them weighed. Remember to take the weight as dressed for the flight. With heavy winter clothing this weight could vary significantly from that taken in a doctor's office. Add the weight of the full fuel and crewmembers to the empty weight of the airplane. If the airplane is experimental or modified from the configuration when it was originally weighed, an actual weighing of the airplane must be performed to determine empty weight. If your selected standard weight is higher than the total of the empty weight, fuel weight, and crew weight, you can either carry additional passengers or some sort of ballast. Lead weights, sandbags, or even bundles of magazines or newspapers will suffice. Remember to tie them down and place them so that the CG is within limits.

During the flight, fuel will be consumed and the weight will be reduced somewhat. After landing, refuel the airplane and note the amount of fuel consumed (Fig. 2-7). Multiply the gallons of fuel used

Fig. 2-7. Monitoring refueling operations after a flight enables accurate fuel consumption measurement to determine average weight during the flight tests.

by 6 to get fuel in pounds. Then take half of this value and subtract it from your takeoff weight to get an *average* value of gross weight for the tests. Let's consider an example.

Suppose your airplane has an empty weight of 1400 lb and a maximum gross weight of 2500 lb. You have decided that you normally operate at about 2400 lb and that you want to evaluate performance at this weight. This fuel capacity is 50 gallons usable and you require a crew of three whose combined weight is 510 lbs. Your gross weight would be computed as follows:

Empty weight:	1400 lb
Fuel weight (50 gal):	300
Crew weight:	510
Total	2210 lb

To bring this figure up to your standard weight of 2400 lb, you add a passenger who weighs 160 lb and 30 lb of ballast.

Now, let's assume that you flew for 1 hour and on refueling the airplane required 10 gal of fuel or 60 lb. The average operating weight would be 2400 lb minus half of 60 or 2370 lb:

$$W_{av} = 2400 - \frac{60}{2} = 2370 \text{ lb}$$

This figure would then be the exact weight to use in your calculations. If you knew that the fuel consumption was about 10 gph and that your flight would take about 1 hour, you could have loaded to 2430 lb so that the average weight would come out to right about 2400 lb. However, 30 lb is only slightly over 1 percent of 2400 lb and should make little difference in performance. Of course, you could choose the maximum gross weight of 2500 lb as standard and load the airplane to that weight to begin your tests. Since you should not ever exceed the maximum gross weight, your average operating weight would always have to be somewhat less than this value.

Weather for Flight Testing

All pilots are aware of the fact that weather conditions impose severe restrictions on flying. Adequate ceilings and visibility must exist throughout the entire flight if it is to be made under VFR conditions. Even IFR operations are not possible if clouds are too low at the destination or if such demons as thunderstorms and icing conditions are present. Flight testing is, probably, the flight operation most sensitive to weather conditions. Not only must the weather be well above VFR minimums for adequate visibility, but it must also be free of turbulence. Turbulence, or the presence of vertical currents in the air, is the biggest obstacle to obtaining good test data.

Turbulence arises from two primary sources. One is the existence of convective currents in the atmosphere. The other is strong wind blowing over uneven terrain. The first of these causes of turbulence, convective currents, can be present even under very calm weather conditions. Convective currents are vertical movements of the air caused by uneven heating of the surface of the earth. Since the heat-reflecting quality of the surface is variable, depending on the nature of the surface material, convective currents vary considerably from one point to another. Green fields and forests reflect relatively little heat, while sandy, dry soil or paved areas reflect considerable heat. For every rising current over these high-heating surfaces there is a compensating downward current over the cooler areas. This situation gives rise to vertical motion of the air, both upward and downward. As an airplane flies through these currents, the alternating upward and downward motion is felt as bumpiness to the occupants. The more severe the vertical currents, the rougher the ride will seem.

Good flight test data require holding the airplane at very precise speeds and attitudes for almost all of the experiments described. Obviously, such procedure is not possible if the air is rough. Furthermore, many tests involve a determination of the airplane's performance in a vertical direction. Rate-of-climb tests and rate-of-descent tests are good examples. If vertical currents are present in the atmosphere, false readings of the airplane's natural climb or descent capability are bound to occur. Sailplane pilots even take advantage of vertical currents to give climb capability to their aircraft, which have absolutely none in calm air. Therefore, certain tests, such as those mentioned, are even more sensitive to atmospheric conditions than others. All tests, however, require relatively smooth air.

Since convective currents, or *thermals,* as they are often referred to by pilots, require heating of the air, they are less severe on cool days. To be more accurate, they are less severe when there is less change in the surface temperature. If the sun comes out very bright and direct after a very cool night, the surface temperature tends to change rapidly and convective currents arise rather rapidly. When the sun is shielded by an overcast, and, in particular, after a fairly warm night, little temperature change occurs to the surface and little thermal activity results. It also takes some time for the surface to heat up, even on bright days. Therefore, early morning is often a good time to run tests. Another good time is just before sunset in the evening, since thermal activity usually subsides at this time. Tests could actually be done at night. However, visual clues and recording of data would be somewhat complicated. Night flying also requires a bit more alertness to factors outside of the airplane. Thus, even though very smooth air can be found at night, it is not recommended as a good time for conducting flight tests.

If testing is done in the morning, it is a good idea to first perform any tests that require low altitudes. This is because the effects of convective currents are lowest early in the morning. As the day progresses and the surface becomes warmer, the turbulence level will rise. The rate-of-climb test, for example, which requires testing at various altitudes, should be started at the lowest altitude. Then, the next higher altitude can be run and so on. In this way you can often keep above the level of thermal activity, even though it does build up.

One way to determine the height of the turbulent air is to note the level of cloud formation. As a connective current rises it cools down due to the cooler air through which it is rising. When it cools to the temperature of the surrounding air, it will not go any higher. If any moisture was present in the rising air, it will condense out and form a cloud when sufficiently cooled. A scattered deck of cumulus clouds, therefore, is a good indicator of the upper boundary of the turbulent air. When sufficient moisture is not present to form clouds, often dust particles or other fine material are carried up by the vertical currents and form a haze layer. The top of the haze layer is, therefore, another good clue to the location of smooth air (Fig. 2-8).

The other major cause of turbulence is strong wind blowing over mountains or other obstructions (Fig. 2-9). An upward motion is caused on the upwind side of a mountain and a downward motion on the downwind side. In addition, a turbulent wake is formed behind

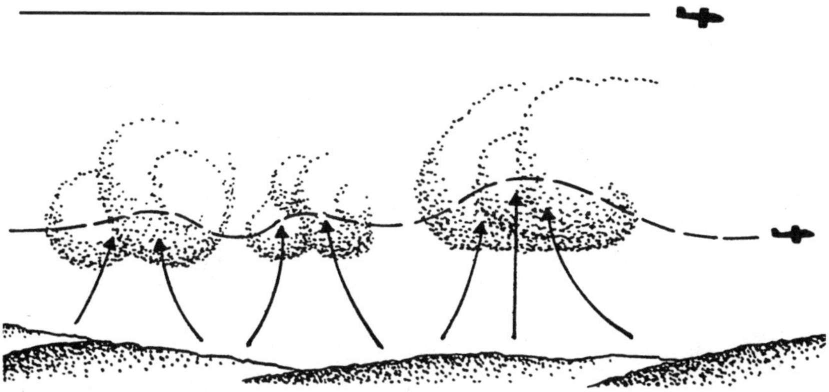

Fig. 2-8. Thermal activity shown ending at cloud level with smooth air above.

the mountain and rather unpredictable air currents are formed. The effects of such turbulence are much like those due to convective currents. The air is rough and makes the holding of accurate conditions difficult. The upward and downward motion also imposes unnatural climbing and descent forces on the airplane. Wind-induced turbulence can sometimes be avoided by climbing above it. However, on days when the wind is very strong, the effects may be felt even at very high altitudes. If low-altitude tests are required, they may not be possible at any time of the day if strong winds persist. If you must fly with some amount of wind blowing, make an effort to avoid mountains or uneven terrain if at all possible.

Another effect that wind can have is to form into very long waves. Such waves may not impart a very rough ride to aircraft flying through it, since the upward and downward currents are widely spaced. However, strong, long wave currents can have significant vertical components to their motion. The airplane could, therefore, be rising or descending due to such wind currents, even though very little roughness in the air is being experienced. It is a good idea to check the consistency of data taken on days when such wind currents could be present. If very unreasonable climbs and descents are being measured, wave effects may be the reason. It is best to cancel the flight tests on such a day and just enjoy the ride.

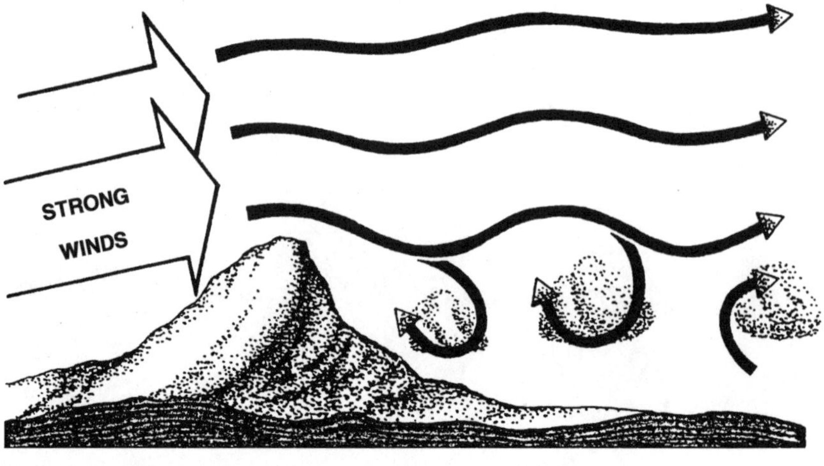

Fig. 2-9. Turbulence caused by mountain wave activity.

Use of Equations

A number of mathematical calculations are necessary to determine performance from the tests outlined in this book. Most of these can be done "by hand," which means by pencil and paper and the standard methods of multiplying, dividing, adding, and subtracting that you learned in school. A simple electronic calculator with only the basic functions would be of tremendous help, however. The use of such a device is highly recommended.

One type of calculator that is particularly useful in performing the calculations required in the following tests is the electronic flight computer. This device, examples are the Sporty's E6B Electronic Flight Computer (Fig. 2-10) or the Jeppesen Sanderson TechStar®, electronically performs all of the functions of the E6B type flight computer, in addition to the arithmetic functions of a standard calculator. It is extremely helpful in determining density altitude and true airspeed under various atmospheric conditions. It can also very easily convert knots to miles per hour or degrees Celsius to degrees Fahrenheit and perform many other special aeronautical functions. In addition, however, it can be used as a regular calculator to work out the equations which will be necessary to reduce the test data.

Fig. 2-10. Sporty's E6B Electronic Computer, an excellent device for quickly and accurately calculating density altitude, true airspeed, etc., as well as general calculations *(Courtesy of Sporty's Pilot Shop.)*

Regardless of the method of calculation, you must have an understanding of the meaning of the equations and the way in which they are set up. If it has been a while since algebra class or, if all of your time there was spent observing the little red-headed girl in front of you, then the following review may be helpful. If, on the other hand, you are a math whiz, then you can skip this section.

Most of us are familiar with the notations for addition and subtraction. $A + B - C$, for example, simply means to add the quantity A to quantity B and subtract quantity C. Multiplication is indicated when two quantities appear side by side, such as AB. This means A multiplied times B. The same meaning is implied if one or both quantities is surrounded by parentheses, such as $(A)(B)$. A divided by B is indicated by placing A over B, as in fractions. A slash means the same thing, such as A/B. Such a term is also sometimes referred to as a *ratio* of A to B.

Higher powers are indicated by superscript numbers. A^2 means A squared or A multiplied times itself. The square root of A is indicated as $\sqrt{A}$ which means a number which, when multiplied by itself, results in A.

Most equations involve combinations of these operations. Brackets and parentheses are used to show the grouping of quantities and the sequence of operations. Consider the following examples of an equation to determine a quantity P:

$$P = \left[\frac{A + (B + C)^2}{D}\right]\sqrt{E}$$

This equation says to add B to C, square this sum, add it to A, divide that amount by D, and then multiply the result (everything inside the brackets) by the square root of E. Suppose $A = 2$, $B = 1$, $C = 3$, $D = 6$, and $E = 25$. The equation would now be set up as:

$$P = \left[\frac{2 + (1 + 3)^2}{6}\right]\sqrt{25}$$

Following the above sequence of operations, the quantities 1 and 3 are first added, and then squared, reducing the equation to:

$$P = \left[\frac{2 + 16}{6}\right]\sqrt{25}$$

Further adding 2 to 16 and dividing by 6 results in the quantity 3 inside the brackets. The square root of 25 is 5, which, when multiplied by 3 gives us our final answer of 15:

$$P = [3]\ 5 = 15$$

Subscripts on characters refer to specific applications of that general quantity. V_S, is used to mean velocity at stall or stall velocity.

W_s is used to indicate standard weight and ρ_0 is used to mean the density at standard sea level. If the subscript involves several words or terms it may be enclosed in parentheses. Parentheses, in this case, do not mean multiplication.

The Greek letter delta, written Δ, is used to indicate an increment or difference in two values of a quantity. For example, Δh means an increment of altitude. We may speak of Δh of 100 ft, or we may say that the Δh between 7200 and 7300 ft is 100 ft.

Occasionally, we use more than one letter to indicate one quantity, such as TAS for true airspeed or GS for ground speed. Most such terms are abbreviations of the words and are, usually, fairly common terms to pilots. Don't confuse them with terms that are to be multiplied. TAS does *not* mean T times A times S, which it would if each of these letters stood for a separate quantity. Note that V_t is also sometimes used to indicate true airspeed (velocity), as is V with other subscripts for other airspeed values. In App. A are standard symbols and subscripts used throughout the book. If in doubt about the meaning of a symbol this table should be consulted.

Presentation of Data

The whole purpose of running performance flight tests is to obtain information on how the airplane performs under various conditions. In order for this information to be of any use, it must be presented in usable form. One of the best ways to present technical data is in the form of a *line graph.* Line graphs are usually constructed from a series of measurements at discrete points which are plotted and then connected by a smooth line—hence, the name line graph. The line may be straight or have some curvature to it, depending on the relationship.

Figure 2-11 shows a graph of runway distance versus time as measured in a takeoff acceleration test. The points represent actual measurements of the distance from the starting point for various amounts of time measured. At 8 s, for example, the airplane was 200 ft from its starting point. The actual points that were measured and recorded are circled. A smooth curve is fitted to these points, however, so that the distance corresponding to any amount of time can be read from this graph. For example, at 11 s the airplane is about 370 ft down

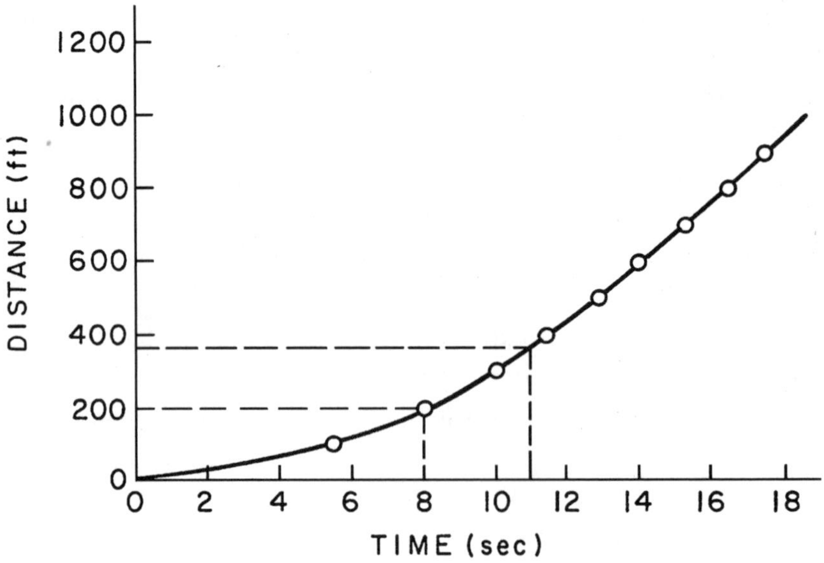

Fig. 2-11. Graph of runway distance versus time in takeoff run.

the runway. Notice that no actual measured point exists, here, however. The graph provides a way of interpolating between actual measurements without additional calculation.

In order to construct the graph in Fig. 2-11, the data had to be measured first. Then, the corresponding points have to be set up in tabular form, such as in Table 2-1.

Table 2-1. Takeoff Runway Distance versus Time

DISTANCES, FT	TIME, S
100	5.5
200	8.0
300	10.0
400	11.4
500	12.9
600	14.0
700	13.3
800	16.5
900	17.6

Next, a suitable scale has to be chosen so that the data will fit the size of the chart. Data must not exceed the boundaries of the chart, yet the scale must be large enough so that the information is readable from the graph. Always mark what each scale represents both in type of information and the units in which it is presented. Here the vertical scale is distance measured in feet and the horizontal scale is time measured in seconds. Usually a graph is referred to by mentioning the vertical scale first, such as a distance-time graph or a distance versus time graph.

In the above example, all of the points fell pretty much on a smooth line. This may not always be the case with experimentally measured data. Sometimes unavoidable error in measurement or disturbances in the atmosphere will yield points that completely defy anyone to fit a smooth curve to them. Usually, such a situation indicates a need for more measurements. Figure 2-12 shows rate-of-climb data measured on a slightly turbulent day. A smooth curve is difficult to construct through these points. Additional tests are conducted on a smoother day and plotted in the squares on Fig. 2-13. Since these points are considered more reliable, they are used to construct the curve along with some of the original points (noted in the circles). Notice that some of the original points, those at 70, 80, and 100 mph, are considered so extreme that they are ignored completely in constructing the curve. All of the other points for a given airspeed are pretty much averaged to estimate the true value of rate of climb at that speed. It is essential that you have some idea of the general shape of the curve expected. Typical curves for various relationships are shown throughout the descriptions of the tests.

The above discussion involved plotting the measured data against other measured data. Under such conditions exact measurements are not possible and significant errors will make the plotting of smooth

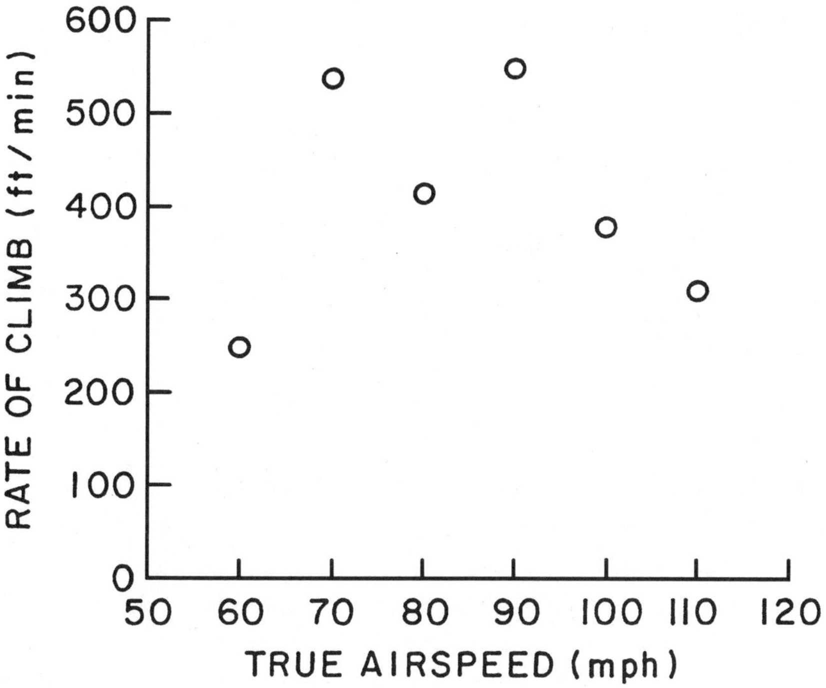

Fig. 2-12. Graph of rate of climb versus airspeed. Points measured on turbulent day.

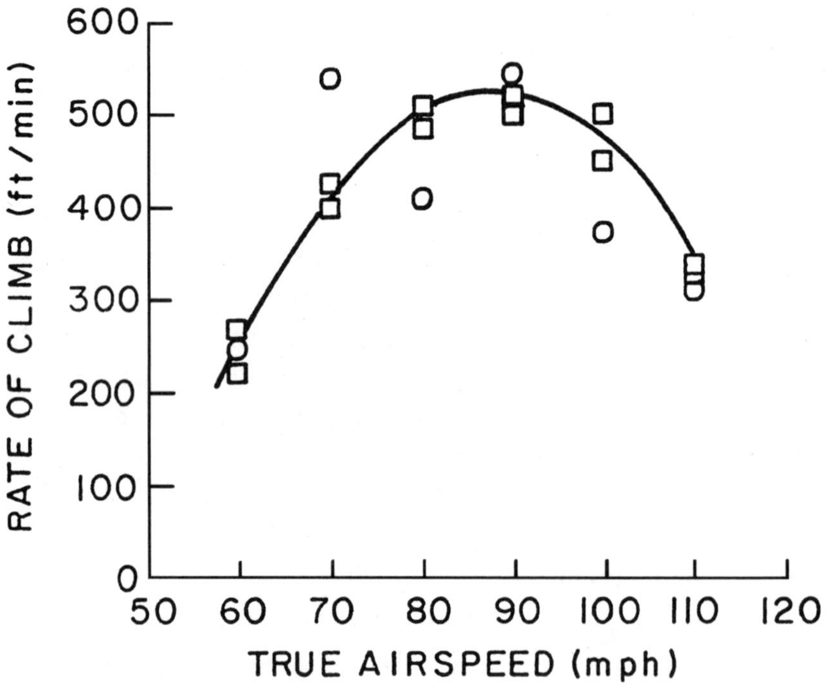

Fig. 2-13. Additional points on a calmer day result in a smoother curve.

curves difficult. Another way of generating curves, however, is from a mathematical relationship, if one exists. Plotting curves from equations requires much less imagination. Once a standard value is known, values under other conditions can often be calculated. For example, in Chap. 4, an equation is given to determine stall speed for any weight, once it is established for one weight. This equation is

$$V_{S_2} = V_{S1}\sqrt{\frac{W_2}{W_1}}$$

Suppose that we have measured the stall speed at 2600 lb and found it to be 65 mph. Stall speed for any other gross weight can then be calculated and plotted as in Fig. 2-14. Calculate as many points as you need to plot a smooth cure. They should all fall right on the line.

Some people find graphs a little confusing and prefer to present data in tabular form. Actually, a table first had to be devised in order to plot Fig. 2-14. The same data as shown on this graph could be presented as in Table 2-2.

Tables are easy to read and yield accurate values as long as you want information corresponding to an exact quantity listed. However, if you require values between these figures, some interpolation is required which could be rather complicated. For example, from the table one could easily note the stall speed for 1800 lb as 54 mph. However, some interpolation is necessary to determine that it would be about 55.5 mph at 1900 lb since 1900 lbs is not listed. It would be even more difficult to obtain a value for 1950 lb. The graph of this same information would yield these values directly, although with some loss in accuracy. The accuracy of information given in graph form depends on the scale at which it is plotted. It does, however, provide for an automatic interpolation process.

Appendix B contains typical performance data as presented in flight handbooks. Graphs of data are given for the Piper Archer and tables are shown for the Cessna 152. It would be useful to study these to see how such information is presented and to determine which form of presentation is to your liking.

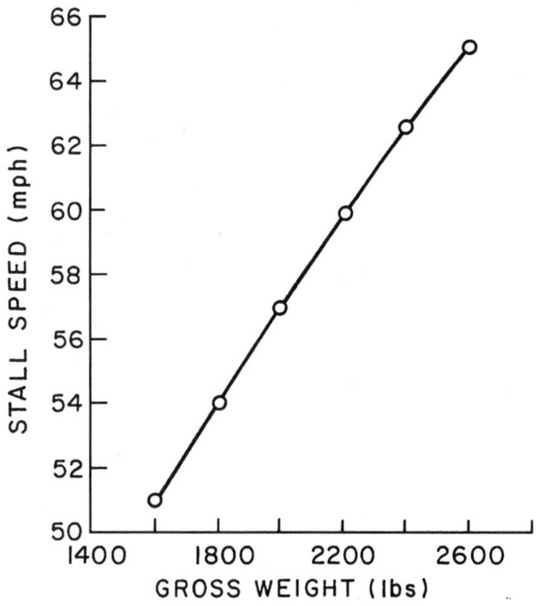

Fig. 2-14. Graph of stalling speed versus gross weight.

Table 2-2. Stall Speed at Various Gross Weights

WEIGHT, LB	STALL SPEED, MPH
2600	65
2400	63
2200	60
2000	57
1800	54
1600	51

CHAPTER 3

Airspeed Calibration

Practically all airplane performance depends on the airspeed at which the airplane is operating. It is, therefore, necessary to know the true airspeed at which each flight test is performed and to know it rather accurately. The airspeed system may have some deficiencies which cause the indicated airspeed to contain some error. Even though this error may be slight, it is important to determine it and correct the indicated values to what they would be if the airspeed system were perfect. This is known as calibrating the airspeed system and involves establishing a chart (or table) to correct indicated airspeed (IAS) to calibrated airspeed (CAS).

Airspeed Errors

Before going into the actual procedure, let's first discuss the reasons for airspeed indicator inaccuracy. There are two basic types of error. The first is *gauge error.* This is the error inherent in the instrument itself and will be the same regardless of where the instrument is installed. All instruments of every type have inherent error to some degree. Modern airspeed indicators, as is the case with most aircraft instruments, have very little gauge error. This is fortunate because this error requires special instrumentation for accurate measurement.

Gauge error can be determined by use of a device which puts pressure into the pitot tube and also into a very accurate pressure indicator. The two values can then be compared. The airspeed indicator is really a pressure gauge itself, but it is marked in values of airspeed rather than pressure readings. Such procedure is shown in Fig. 3-1. This method of checking gauge error should be done only by someone familiar with both the equipment and the procedure. It is quite easy to damage the airspeed indicator by putting too sudden or too much pressure into it. You probably recall your flight instructor telling you never to blow into the pitot tube. This is why. If you suspect a significant gauge error, it is wise to have it checked by a qualified instrument repair shop.

The second type of error is called *position error.* This error is due to the location of the pitot and static ports in the pressure field surrounding the airplane. This error will be different for the same indicator installed in different airplanes. The pitot tube is aimed directly into the air stream and picks up the total pressure; that is, the dynamic pressure, or that pressure from the moving air plus the pressure of the surrounding still air. The static port is meant to measure only the static or still air pressure. The airspeed indicator operates by differentiating between these two pressures and reading

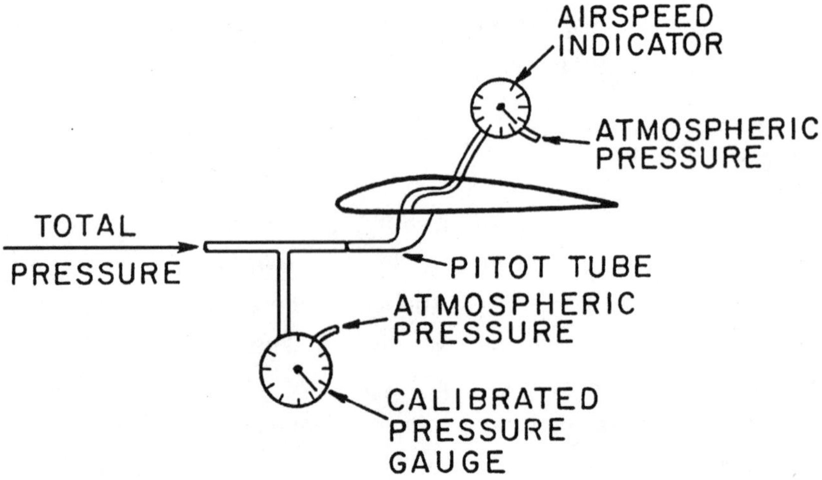

Fig. 3-1. Gauge error on test setup.

out only the dynamic pressure. Obviously if the pressure at the static port is not quite the true static pressure then the difference the indicator measures will be wrong and an airspeed error exists.

Pitot tube and static port locations are designed with considerable care to minimize error. However, the error usually will not be consistent, since any change in attitude of the airplane causes a change in the pressure field surrounding it. The pitot tube is usually located under the wing because here the flow is nearly always aligned with the wing surface regardless of the angle or attack. Figure 3-2 shows such an installation. This method of mounting usually all but eliminates error in the pitot tube.

The static port is harder to locate. It sometimes is located with the pitot tube or somewhere on the aft portion of the fuselage. It may incorporate some sort of baffle to minimize the effect of change in air direction. However, as the angle of attack changes slightly, it may pick up a little effect from dynamic pressure. This amount can vary with various angles of attack. Figure 3-3 shows a typical static port on the fuselage.

Fig. 3-2. Typical pitot tube location beneath wing.

Fig. 3-3. Typical static port location on side of fuselage.

Test Procedure

Because the airspeed is dependent upon angle of attack, position error will probably vary with different airspeeds. Therefore, it is necessary to check the error for a number of airspeed values throughout the entire range normally flown. Position error alone is difficult to determine accurately without knowing the exact gauge error. The entire system error, however, which includes both gauge and position error, can be determined rather easily.

Various flight test methods are employed by the aircraft industry and research organizations to accurately calibrate airspeed. One of the more popular is the photogrid method, in which the airplane is flown by a tower containing a motion picture camera shooting through a calibrated grid. The aircraft's progress across the grid is recorded along with the elapsed time, so that groundspeed can be determined. Accurate wind information is also measured, and applied to correct the speed to true airspeed. Another method is the pacer method, where the airplane to be calibrated is flown in formation with another containing an airspeed system of known calibration. Indicated airspeeds are varied and compared. Yet another method involves towing a body, whose aerodynamic characteristics are quite predictable, outside of the field of influence of the airplane. The static pressure, or both static and total pressure, on this body is measured to determine accurate airspeed, which can be compared to that read in the panel indicator. This method is often called the "trailing bomb" method.

All of these methods involve special equipment, some of which is rather sophisticated (translated, expensive). Another quite legitimate method is sometimes used, though, which does not require any special instrumentation. It is known as the speed-course method, and consists of flying over a calibrated range, or distance, and recording the time that it takes to cross it. This method is probably best suited to the type of flight testing described in this book.

The first step is to select a calibration range. This means choosing two straight, parallel landmarks a known distance apart. Roads, railroads, fences, power lines, or any straight, prominent objects can be used. If you don't know the distance between the objects, measure it. A calibration range should be about one to five miles in length, though it doesn't have to be an even number of miles. Just be sure you know the *exact* distance. Figure 3-4 shows a typical range. The fence line and the highway parallel to it form two boundaries 3.5 miles apart. Notice that the road running perpendicular to these landmarks crosses both of them and therefore can be used to measure the distance. This is best done with a tape measure but can be approximated by using an automobile with an accurate odometer. Quite often an airport will have a known calibration range nearby, so ask your FBO or some of the old-time pilots around the field. This will save you a lot of effort, because the biggest part of this procedure is selecting and measuring a test range.

The only other equipment you need is a stopwatch. The basic procedure simply is to fly across the range at a constant airspeed and record the time it takes to go this known distance. This would be the complete test if there were no wind. The true airspeed would equal the ground speed which would be the distance divided by the time. It could be compared with the airspeed value read from the indicator

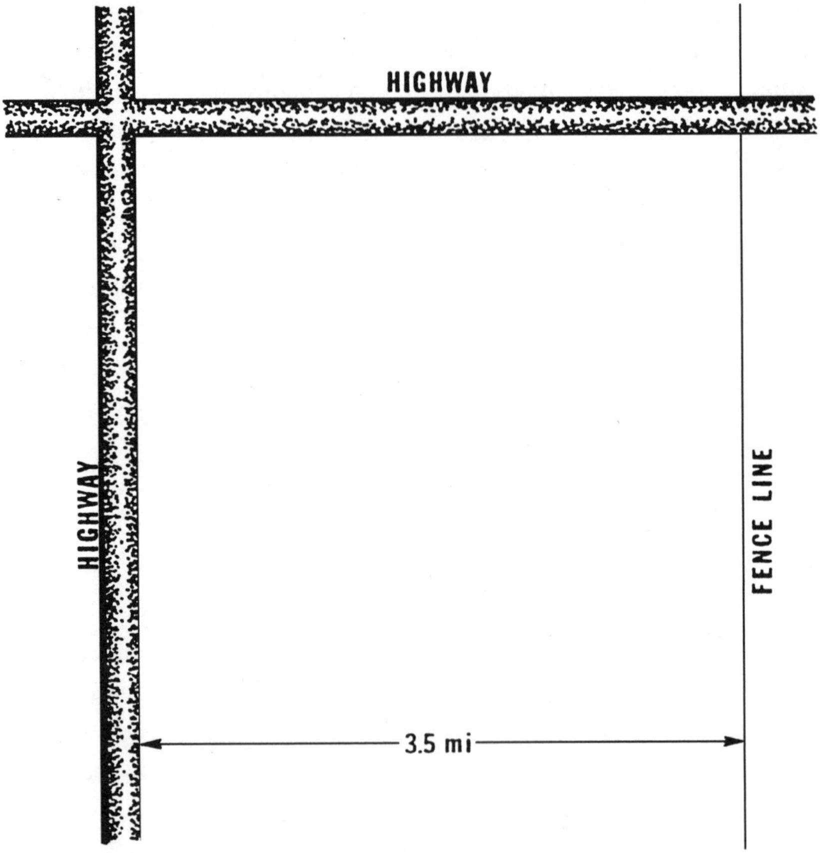

Fig. 3-4. Typical airspeed calibration range.

(corrected to TAS) and the error determined. However, as all pilots well know, wind nearly always is present and usually is the hardest factor to determine accurately in time-speed-distance problems. Well, don't give up yet, because the wind problem can be solved very easily for our purposes here.

Suppose we had a direct headwind over our course. The wind would subtract from our groundspeed by the amount of the wind speed. Now, suppose we turned around and flew the course in the opposite direction. The wind would now add to our groundspeed by this same amount. Thus, if we flew the course in both directions, and averaged our ground speed, we would get the true airspeed. For example, suppose the airplane had a true airspeed of 100 mph and the wind were

20 mph. Flying from landmark *A* to landmark *B* the groundspeed would be TAS minus the windspeed or 80 mph. Flying in the opposite direction, the wind would create a tailwind and the groundspeed would be 100 mph plus the windspeed or 120 mph. Now if we average these two groundspeeds, that is, add them and divide by two, we end up with the original true airspeed of 100 mph. Therefore, if we can measure the groundspeed in both directions, we can calculate the true airspeed. Groundspeed is easily determined as the distance divided by the time to go that distance. This method is shown in Fig. 3-5. The procedure is:

1. Calculate groundspeed from *A* to *B*.

$$GS(A \text{ to } B) = \frac{\text{distance in miles}}{\text{time in minutes}} (60)$$

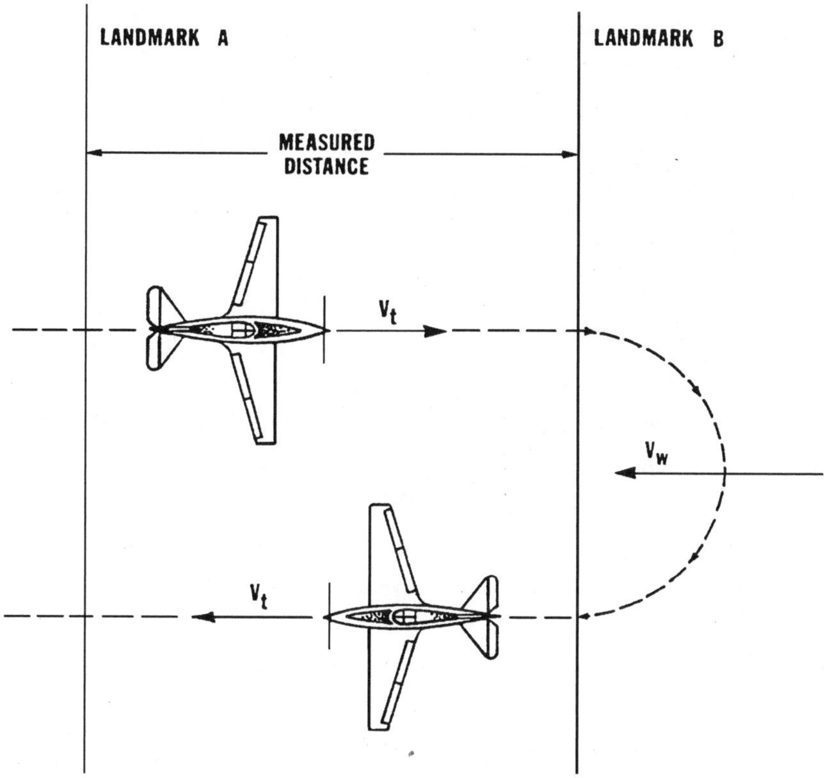

Fig. 3-5. Airspeed test procedure with direct headwind and tailwind.

2. Calculate groundspeed from B to A.

$$\mathrm{GS}(B \text{ to } A) = \frac{\text{distance in miles}}{\text{time in minutes}} (60)$$

3. Calculate the average groundspeed by adding the two groundspeeds and dividing by 2.

$$\frac{\mathrm{GS}(A \text{ to } B) + \mathrm{GS}(B \text{ to } A)}{2}$$

Steps 1 and 2 can also be performed by use of a flight computer.

Now, let us consider the case where we have a wind with a crosswind component as in Fig. 3-6. If we keep the airplane headed in the direction *a-b,* just as we did for the no-wind or direct headwind-tailwind case, and neglect the drift, we will keep the true velocity vector headed in the proper direction. The only part of the wind that will affect the groundspeed will be the headwind-tailwind component as indicated on the figure. The drift component will not affect our groundspeed since it is perpendicular to our heading. It will have the effect of causing the airplane to drift to that it does not actually arrive at point *b* but rather at point *c.* This position is of no consequence to the problem, however. The procedure, therefore, is exactly the same as before.

If you don't follow all of the mathematics, the procedure is very simple. Just head the airplane perpendicular to your landmark lines, neglect the drift, and fly a constant heading, constant airspeed course, recording the time it takes to fly from one line to the other. Then fly a reciprocal heading at the same indicted airspeed and again record the time. Calculate each groundspeed by dividing the distance by the time, add the two groundspeeds, and divide by two. The result is your true airspeed. Now, change your measured true airspeed value to calibrated airspeed. This can be done with your flight computer in just the same way you make TAS corrections, except that you read a calibrated value from the inner scale for a given value of true airspeed on the outer scale. This calibrated value can then be compared with your indicated value read from the gauge to determine your error. If you do this for a number of different indicated airspeeds, you could

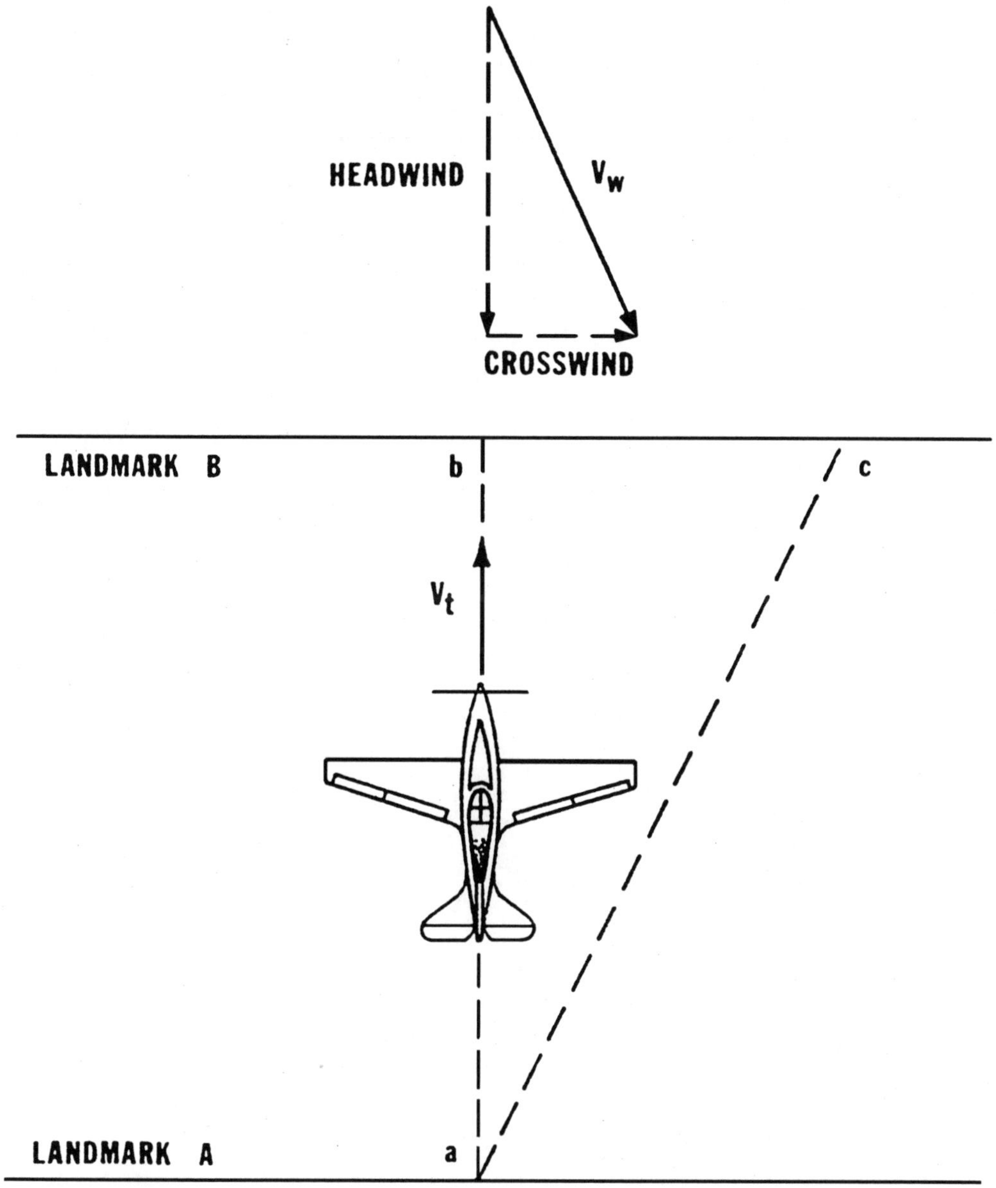

Fig. 3-6. Airspeed test procedure with crosswind.

then plot a chart of indicated versus calibrated airspeed. Figure 3-7 is such a chart, and it shows, for example, that for an indicated value of 85 mph your calibrated or actual value is only 80 mph. This chart resulted from actual flight tests by the author on a modified Globe Swift.

It usually is best to utilize two people in conducting the tests. In this way the pilot can concentrate on holding a constant heading and airspeed while the observer can sight the landmarks and operate the stopwatch. The observer should sight over some fixed point on the aircraft, such as a point on the wing leading edge for a

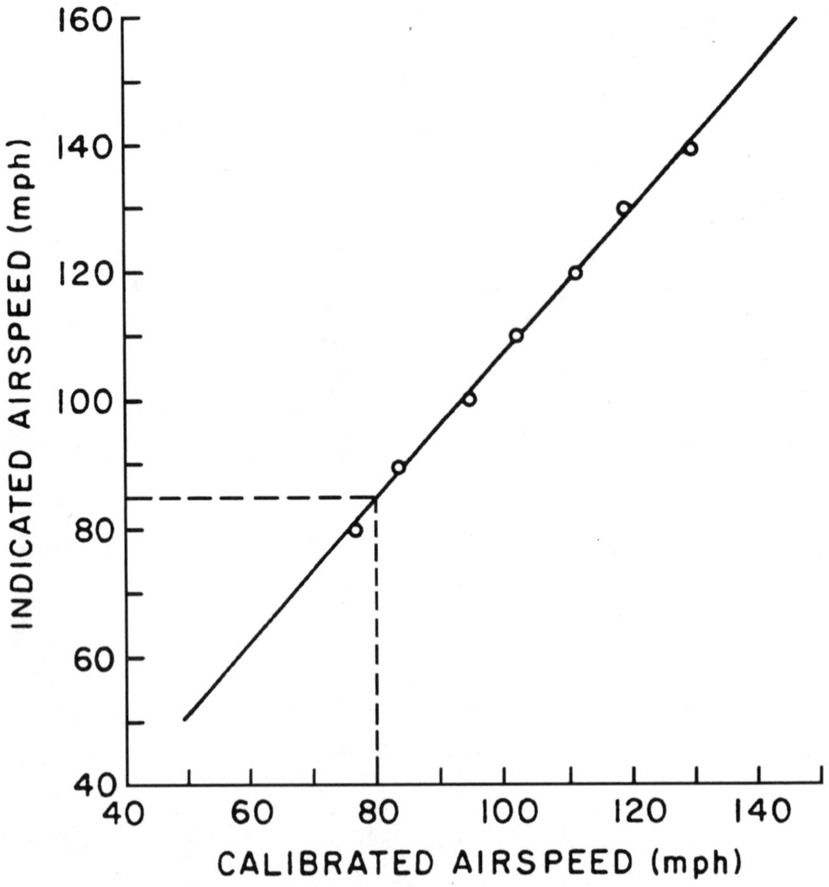

Fig. 3-7. Airspeed calibration chart as obtained from flight tests of modified Globe Swift.

low-wing (Fig. 3-8) or a point on the strut or landing gear for a high-wing (Fig. 3-9). It is advisable to start the run some distance from the first mark so that the airplane can be trimmed out, and its power set up for a constant speed and altitude, before the timing is started. A constant altitude is desirable during the run but slight changes such as a hundred feet or so will not alter your measurements appreciably. Make your runs as low as regulations or safety precautions permit for the best accuracy in timing. Of course, a very low amount of atmospheric turbulence is a must for any degree of accuracy.

GPS Method

The Global Positioning System (GPS) has revolutionized navigation, both in the air, and in other modes of travel. Its high degree of accuracy has many other applications, however. One of its features that can be employed in airspeed calibration is a quite accurate measurement of groundspeed. Although we can calculate this from timing the

Fig. 3-8. Using wing leading edge as a guide to crossing airspeed range landmarks with low-wing.

Fig. 3-9. Using wheel fairing with high-wing airplane as guide to crossing landmark.

distance over a measured range, the GPS gives us a direct reading of groundspeed at any time. We still need to average the groundspeed to get true airspeed, but obtaining the groundspeeds involved is much easier, quicker, and more accurate with GPS. In fact, the basic procedure is so simple that it is becoming common to fly in opposite directions, and average the groundspeed readings from each run to determine an accurate airspeed value.

The exact procedure is not quite that simple, however. As with the calibration range method, the heading of the airplane makes a difference in the readings. The reason is that winds have a definite effect. Consider the airplane in Fig. 3-6, showing the effects of a crosswind. We stated that the track over the ground is not from point *a* to point *b*, but rather to point *c*. In the calibration range method, this drift did not matter, because we cross landmark *B* at the same time, whether there is a crosswind or not. We just cross it at different points, depending on the strength of the crosswind. The GPS, however, measures our groundspeed *along the ground track,* which is from *a* to *c*. Since this path is longer than that from *a* to *b*, the airplane must be moving faster than it would if there were no crosswind, and it went directly to *b*. The same situation would be true in the return run. Hence, we would have two groundspeeds higher than those resulting if there were no crosswind, and when averaged, would yield a higher true airspeed.

The moral of the story is that we can't tolerate a significant crosswind in measuring true airspeed with the GPS. The saving grace of the GPS method, however, is that you can head in any direction; you are not bound by the heading dictated by the range landmarks. Hence, we head directly into the wind for our first groundspeed reading, and in the reciprocal direction for the second reading. We must first, however, determine exactly what the wind direction is. A starting point would be to consult the winds aloft forecast for the altitude selected. Another neat feature of this method is that we are not bound by any altitude limits, so we can go as high as necessary to reach smooth air, or to obtain favorable wind.

Once you have climbed to the desired altitude, select a mode on the GPS that displays *ground track*. Head the airplane in the direction of the forecast wind (on your heading indictor or compass), and watch the ground track display. If it stays constant, you have an indication that you are not drifting, and are, indeed, heading into the wind. If it changes, correct your heading slightly, and watch again. Continue to do this until you establish a heading that maintains a constant track.

Now you can select various airspeeds, and record the corresponding groundspeed readings on the GPS. Then turn the airplane 180 degrees, and repeat the procedure. There is no need to reverse course with each airspeed value. It is much easier to go through all the data points in one direction, then reverse, and run through them all in the opposite direction. Make sure that you record the pressure altitude (altimeter set to 29.92) and outside air temperature for the tests.

From this point on, the procedure is exactly the same as that for the speed course method. Average the two groundspeed values for each indicated value read on the gauge, by adding them and dividing by two. The result is your true airspeed for that indicated value. What you want, though, is the *calibrated* value, in order to determine the amount of error. Hence, the TAS must be multiplied by the square root of the density ratio to yield CAS. This could also be done by use of an E6B, or an electronic flight computer.

It should be noted that, if the wind is very light, the drift does not affect the true readings significantly. Even in a fairly strong wind, a small amount of drift has little effect. Don't spend too much time trying to get your heading and track to agree exactly. There is also a way of doing this test without heading directly upwind and downwind, and that is to measure the angle of drift. This is the angle between the lines *a-b* and *a-c* on Fig. 3-6. This angle would be the difference between the heading and the track, while allowing the airplane to drift. By trigonometry, you could then compute the component of the groundspeed reading that is acting in the direction of the heading. This procedure is more complicated, though, and not recommended. The direct-into-the-wind method is much easier.

True Airspeed

As mentioned previously, the airspeed indicator is really a pressure gauge which measures the dynamic pressure of the air. This pressure depends on the air density and is equal to $\frac{1}{2}\rho V^2$, that is one-half times the density times the velocity squared. Since the indicator *reads* in values of velocity, however, some constant value of density must be used. The value is standard sea level density. This means that the indicator, even in the absence of any error, never reads true airspeed except when the density is at standard sea level. Thus, to get true airspeed (TAS) we must make a correction for density anytime it is different from that at sea level. Since this is most of the time, calibrated airspeed normally has to be corrected to determine TAS.

It was already mentioned before that corrections between CAS and TAS can be made by use of a flight computer. To do this you must set in the proper pressure altitude and temperature. These are the quantities needed to measure density and what the computer is really doing is calculating the density and its difference from standard sea level density. You can also make this correction without a computer by use of the following equation:

$$\text{TAS} = \frac{\text{CAS}}{\sqrt{\rho/\rho_0}}$$

The term ρ/ρ_0 is the density ratio and can be determined from Figs. 1-3 and 1-4 by using the proper temperature and pressure altitude. Dividing the square root of this ratio into the calibrated airspeed yields the true airspeed.

The normal procedure to determine true airspeed is then done in two steps:

1. First, read a value of IAS from the airspeed indicator and correct it to CAS by the calibration curve established in the test in this chapter.
2. Correct the value of CAS to TAS by applying the temperature and pressure altitude (at the time of the test) to either a flight computer, or to the charts to obtain ρ/ρ_0 and then use the above equation.

There is another airspeed definition that should be mentioned here: equivalent airspeed (EAS). This speed is calibrated airspeed corrected for compressibility. Compressibility occurs when the airplane reaches a high enough Mach number that the air entering the pitot tube starts to compress. This situation becomes significant when the aircraft is going over about 200 knots (kn) or above 10,000-ft altitude. Strictly speaking, the definition of true airspeed is *equivalent* airspeed corrected for density. At the speeds and altitudes for which most lightplane flying is done, however, the difference between CAS and EAS is negligible. Therefore, we normally consider them to be the same, and consider true airspeed to be CAS corrected for density. If you are testing a high-speed aircraft, though, you may need to consult an equivalent airspeed correction chart to account for compressibility.

True airspeed is normally required when determining performance involving the time to cover a certain distance. The time to ETA, for example involves a determination of groundspeed, which, in turn depends on TAS. Thus, cruise performance is usually established for true airspeeds. Takeoff performance also relies on TAS since the time to cover a certain distance on the ground is used to measure the acceleration of the airplane.

On the other hand, certain qualities of the airplane depend on dynamic pressure. Since this is what the airspeed indicator measures, indicated airspeed is really a better determination of such performance. Stall speed, for example, depends on lift, which is dependent on dynamic pressure. The IAS at which the airplane stalls is, thus, the important figure. Climb performance, also, is usually determined for IAS, since this is what the pilot is reading in order to establish a climb speed.

Airspeed Calibration Example (Speed-Course Method)

Conditions:

- Distance between landmarks (A to B) = 3.0 mi
- Time to fly from A to B = 2.3 min
- Time to fly from B to A = 2.1 min
- Temperature = +20°C (68°F)
- Pressure altitude = 1000 ft
- Indicated airspeed = 85 mph

1. First calculate groundspeed from A to B:

$$\text{GS} = \frac{3\ \text{mi}}{2.3\ \text{min}} \times 60 = 78\ \text{mph}$$

2. Second, calculate groundspeed in the opposite direction (B to A):

$$\text{GS} = \frac{3\ \text{mi}}{2.1\ \text{min}} \times 60 = 85.71\ \text{mph}$$

3. Now average the groundspeed:

$$\text{GS(average)} = \frac{78.26 + 85.71}{2} = 81.99\ \text{mph}$$

Round this off to 82 mph.

4. This is your true airspeed. Convert this to calibrated airspeed (CAS). Set 20°C opposite 1000-ft pressure altitude on your flight

computer as shown in Fig. 3-10. Opposite 82-mph TAS on the outer scale read 80-mph CAS on the inner scale (Fig. 3-11).

5. This value of 80 mph is the calibrated airspeed for which you read 85-mph indicated airspeed on the gauge. Plot it and repeat the entire process at other indicated airspeed values.

Airspeed Calibration Example (GPS Method)

First, determine wind direction at altitude flown for the test by correlating heading and track, and fly directly into the wind.

Conditions:

- Pressure altitude = 3300 ft.
- Temperature = 58° F.
- Indicated airspeed = 75 kn.

1. Record groundspeed reading: 80 kn.
2. Turn 180° and read groundspeed: 75 kn.

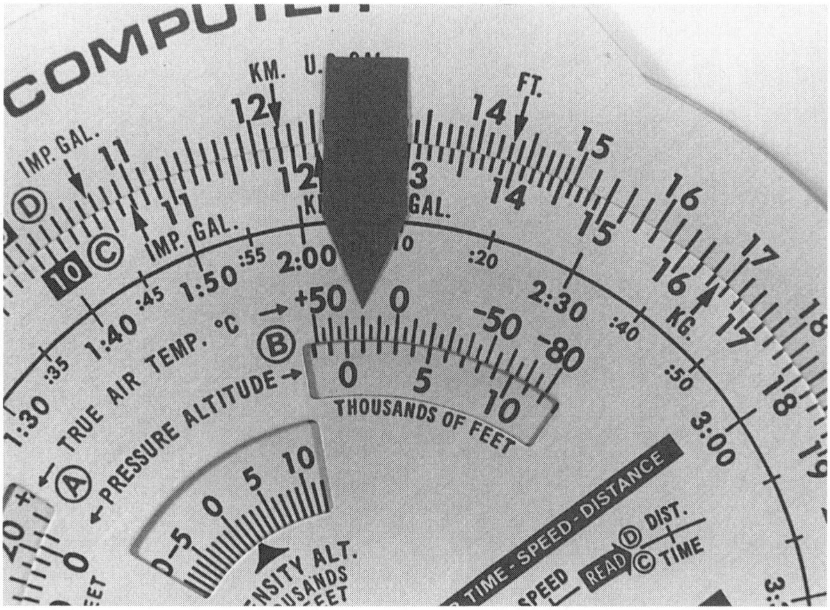

Fig. 3-10. Setting 20°C temperature opposite 1000-ft pressure altitude on flight computer.

Fig. 3-11. Reading 80-mph CAS below 82-mph TAS on flight computer.

3. Average the two groundspeeds:

$$\text{GS(average)} = \frac{80 + 75}{2} = 77.5 \text{ kn}$$

4. This value is TAS. Convert this to calibrated (CAS) by one of the methods described before. The result is 73 kn.

5. This value of 73 kn is then the calibrated airspeed for which you read 75-kn IAS on the gauge.

CHAPTER 4

Stall Speed

Most pilots are aware that lift is created on an airplane wing by a difference in pressure between the top and bottom surfaces. The lower pressure on the top surface is achieved by an application of an old principle known as Bernoulli's theorem. Very simply, this theory states that if the velocity of a moving stream of air is increased, the pressure it exerts on a body over which it flows will be less. The airfoil shape of the wing takes advantage of this principle to lower the pressure on the top of a wing more than on the bottom.

Figure 4-1*a* shows an airfoil at a small angle of attack. At some point near (but not necessarily right at) the leading edge, the flow divides. Part of it goes over the top surface and part over the bottom. Because there is more top surface than bottom, the flow must speed up over the top surface. This is required in order to have a continuous flow of the air and satisfies another principle of science knows as "continuity." This term is just the old law of conservation of mass applied to a moving stream of fluid such as air. When the flow is speeded up, the pressure must be reduced according to the Bernoulli principle.

When the angle of attack increases, as shown in Fig. 4-1*b*, the dividing line moves back on the lower surface. The flow over the top surface now has a longer path and, therefore, must speed up even more. This process lowers the pressure on the top to a greater degree.

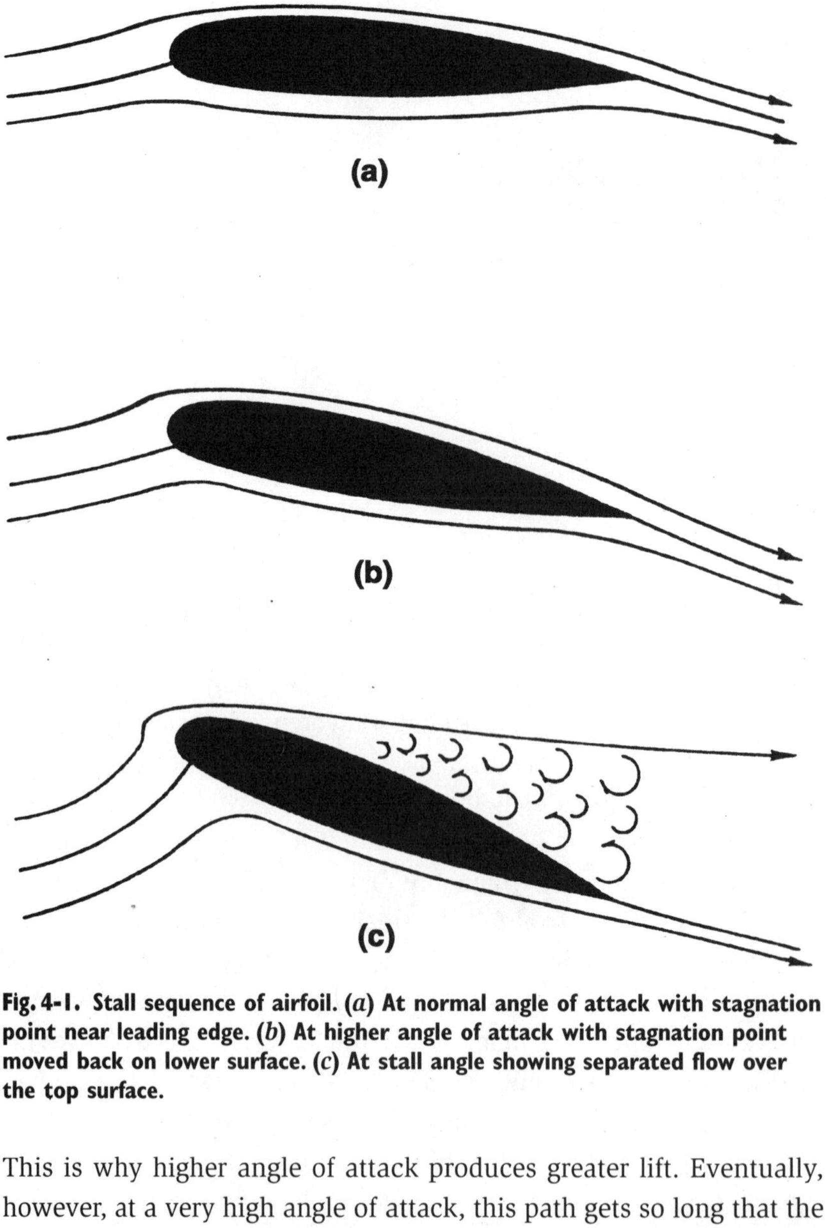

Fig. 4-1. Stall sequence of airfoil. (*a*) At normal angle of attack with stagnation point near leading edge. (*b*) At higher angle of attack with stagnation point moved back on lower surface. (*c*) At stall angle showing separated flow over the top surface.

This is why higher angle of attack produces greater lift. Eventually, however, at a very high angle of attack, this path gets so long that the air doesn't have enough energy to travel this entire path. Remember that it is being retarded by friction along the surface, and the longer the path, the greater this frictional effect will be. At the limit, it breaks away from the top surface. The result is a flow pattern as shown in Fig. 4-1*c*. The flow over the top surface is separated from the leading edge back. With no smooth flow over the top surface there is nothing to lower the pressure and the lift drops off significantly. This condition is what we call a "stall."

Lift does not depend upon angle of attack alone, but is also affected by the density of the air, the velocity, and the wing area. Mathematically the equation is

$$L = K \alpha \rho V^2 S$$

where L = lift
K = constant of proportionality
α = angle of attack
ρ = density
V = velocity
S = wing area

This equation says that it is the product of all of these factors multiplied together, including the square of velocity, that determines lift. The constant, K, is a number depending on the units of the various factors, and also the wing and airfoil shape. At a given altitude (constant density) and with a given wing area, the two remaining variables are angle of attack and velocity. Including the density and wing area in the constant would then make the equation

$$L = K \alpha V^2$$

Thus, to maintain a certain amount of lift we can trade off velocity for angle of attack. This is why at low airspeeds we need to have a high angle of attack, and at high speed, a low angle of attack. Furthermore, from a balance of the four basic forces on the airplane (lift, weight, thrust, and drag), the lift must equal the weight in steady level flight. Therefore, we can say

$$L = W = K \alpha V^2$$

where W is the gross weight of the airplane.

If we plot lift versus angle of attack for a certain airfoil section, it looks like the graph in Fig. 4-2. Notice that the lift goes up proportionally to angle of attack until the stall angle is reached. Again, for a given airfoil there is a *fixed* angle of attack at which it stalls. Therefore,
at the stall point, we can include the angle of attack in our constant and

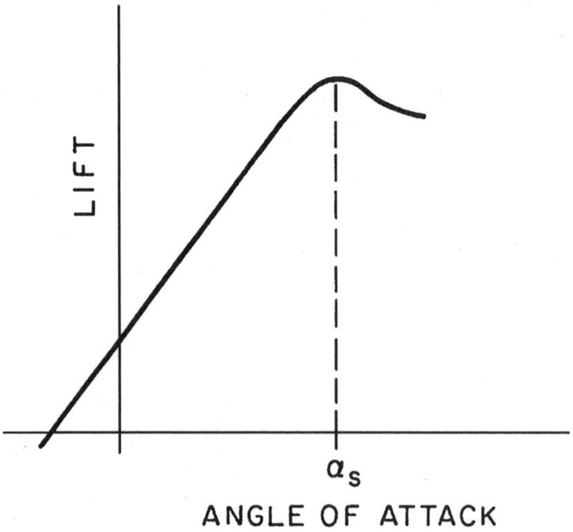

Fig. 4-2. Lift versus angle of attack showing location of stall angle.

the lift equation is

$$L = W = K\,V^2_S$$

This equation shows that the higher the weight, the higher our stall velocity must be. For this reason we need to consider the stall speed at various gross weights.

Another factor that affects stall speed is the flap setting. In our original equation for lift, a factor K was included to account for, among other things, the airfoil shape. Lowering flaps changes the airfoil shape and thus alters K. It turns out that for higher flap deflection angle, this constant becomes higher and thus, more lift is obtained. All of this means that another set of stall speeds will occur for various weights with a certain degree of flap deflection. Hence, to get a complete set of stall data, stall speed must be measured with various angles of flap deflection.

Fortunately, different altitudes do not affect stall speed if we consider only calibrated airspeed (CAS). The *true* airspeed at which an airplane stalls is different at different altitudes for the same weight and flap setting. However, the density change with altitude affects the airspeed indicator proportionally to the way it affects stall speed. If stall speed were measured at sea level, for example, and then several thousand feet above sea level, the lower density at the higher altitude would cause the airplane to stall at a higher TAS. However, the lower density would also reduce the dynamic pressure in the pitot tube and the indicator would read no higher than at sea level (assuming the instrument error to be constant). This is the reason you normally have to calculate TAS from an indicated value.

Bank Effects on Stall

When the airplane is flying straight and level, the lift being generated by the wing is equal to the weight. The lift force is perpendicular to the plane of the wing and, in this case, this plane is parallel to the horizon. However, when the airplane is banked, the lift is still perpendicular to the plane of the wing, but now the wing is tilted with respect to the horizon. The lift force is, therefore, tilted. Weight is still the same and always tends to pull vertically downward. Thus, more lift must be generated so that the *vertical component* of the lift vector equals the weight. In addition, the airplane in a bank is experiencing a centrifugal force due to the turning of the airplane. Part of the lift must be utilized to overcome this centrifugal force. The horizontal component of the lift provides this force.

For these reasons you can see that additional lift is required to maintain level flight in a bank. Stall speed is proportional to the square root of lift, so that greater lift results in higher stall speed. Remember that in level flight, higher gross weight can result in higher stall speed for the same reason. Higher weight requires higher lift to overcome stall speed. For whatever reason, when more lift than normal exists, stall speed will be higher. Therefore when banking is done the stall speed goes up. Steeper bank requires more lift; consequently, higher stall speed will result. The amount of increase with bank angle is the same for all airplanes. Therefore, the stall speed for any bank angle can easily be calculated by use of Fig. 4-3. An airplane that stalls at 50 kn in level flight, for example, would stall at 1.25 at times that speed in a 50° bank, or at 62.5 kn.

Test Procedure

The actual flight testing of stall speed is relatively simple. Since the test depends on weight, it is necessary to determine the gross weight rather accurately. The airplane should then be flown to a safe altitude and area in which to perform stalls. Since calibrated airspeed for stall does not vary with altitude, the exact altitude at which the tests are conducted does not matter.

The stall should be performed power off and entered very gradually. Abrupt entry will result in an accelerated condition and will not yield accurate results. It is best to keep the airplane in fairly level attitude (nose on or slightly above the horizon) and definitely keep

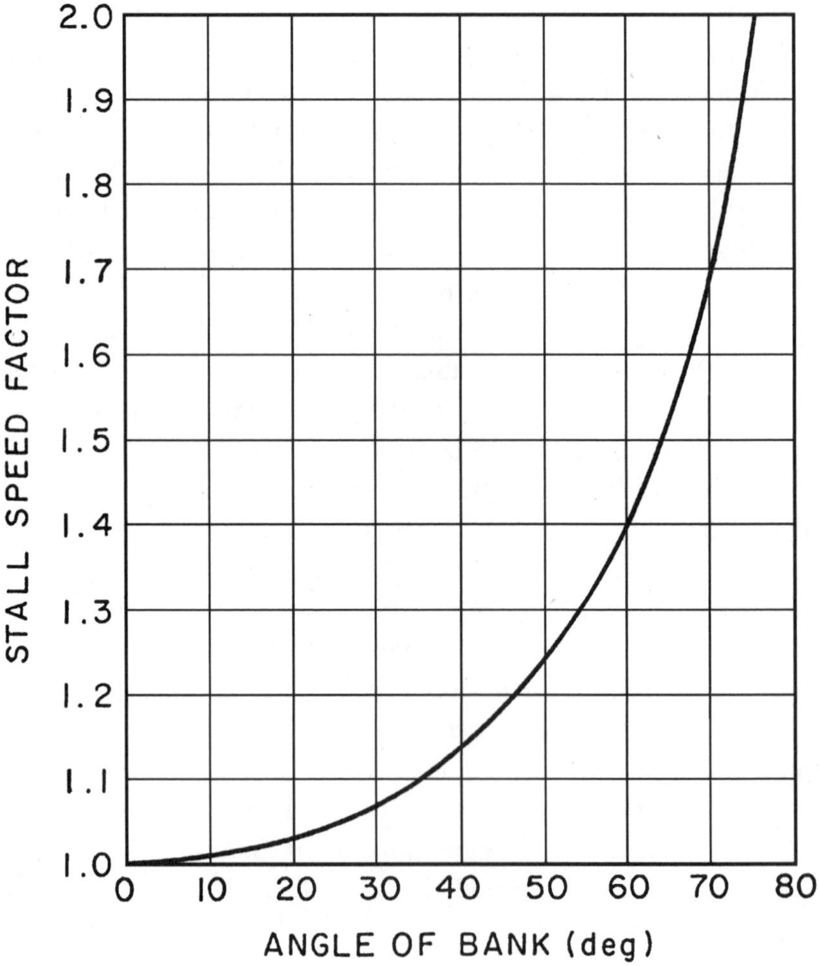

Fig. 4-3. Stall speed factor versus angle of bank.

the wings level (Fig. 4-4). Very slowly pull back on the stick and watch the airspeed indicator. The needle should slowly and smoothly move down and stop just at the break of the stall. Record this velocity. The procedure should then be repeated at various flap settings. If definite notches of flap settings are incorporated in a manual controlled system, such as in many Pipers, measure the stall speed for each setting. If a continuous motor-driven system is employed, as in most Cessnas, measure full deflection and as many other settings as you desire. For a retractable gear airplane, you should also measure stall speed in both gear-up and gear-down configurations. The FAA

Fig. 4-4. Stalling the airplane with nose nearly level.

specifies a stall speed for landing configuration known as V_{S0}. This is the stall speed with engines idling, gear extended, propellers in takeoff position, and flaps in landing configuration. FAR Part 23 requires all single-engine airplanes and certain light twins to have a stall speed in this configuration of no greater than 61 kn (70 mph).

To determine the stall speed at other weights, the gross weight could be changed by changing the loading and the experiment repeated. However, once the stall speed is known for one weight, it can be calculated for other weights by the following equation:

$$V_{S_2} = V_{S1}\sqrt{\frac{W_2}{W_1}}$$

This equation says that if you know the stall speed V_{S_1} for one gross weight W_1, then the stall speed for another weight V_{S_2} can be determined by multiplying V_{S_1} by the square root of the new gross weight W_2 divided by the original gross weight. This same situation holds true for the various flap settings, also.

In any case, the values you obtain are in indicated airspeed (IAS). They will have most meaning to the pilot if plotted that way, that is, IAS versus gross weight for various flap settings. Figure 4-5 shows

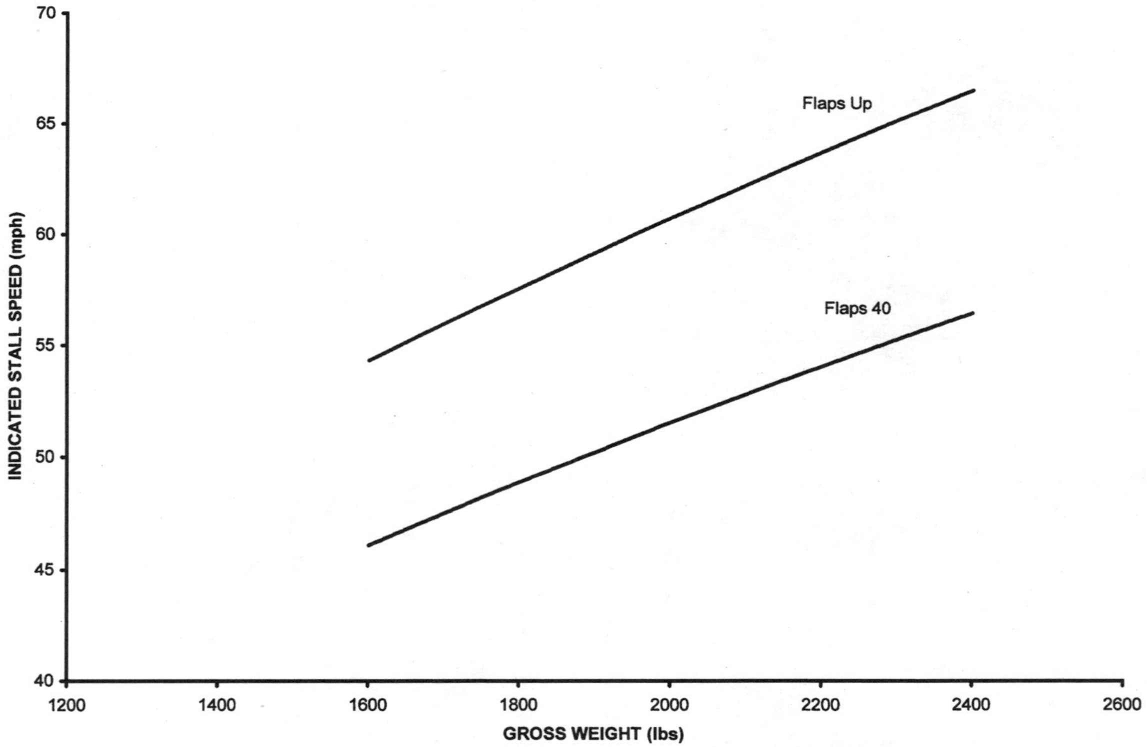

Fig. 4-5. Typical chart of stall speed versus gross weight for various degrees of flap deflection.

such a chart.

If you need the true airspeed value of stall speed, the IAS value must first be corrected to CAS by the chart constructed in Chap. 3 and converted to TAS by correcting for appropriate altitude and temperature. This procedure can be done with the flight computer as discussed also in Chap. 2.

Stall speed for any angle of bank can also be plotted into a chart by taking the appropriate value of stall speed for wings level, and multiplying by the various factors for the respective bank angles from Fig. 4-3.

Stall Speed Test Example

Test Conditions:

- Gross Weight = 2430 lb
- Flaps = 0°
- Stall Speed = 67 mph IAS

1. Determine stall speed at 2500 lb, use the weight correction equation.

$$V_{S2500} = V_{S2430}\sqrt{\frac{2500}{2430}}$$

$$= 67\sqrt{\frac{2500}{2430}}$$

$$= 68 \text{ mph}$$

2. To determine stall speed at 2000 lb, use the same equation.

$$V_{S2000} = 67\sqrt{\frac{2000}{2430}}$$

$$= 61 \text{ mph}$$

Note that the airspeeds are rounded off to the nearest whole mile per hour in these examples.

CHAPTER 5

Takeoff Performance

The whole purpose of a takeoff run is to get the airplane up to flying speed. The airplane will fly at any speed above stall. However, to lift off at just one or two knots above stall is a bit risky. Just the slightest increase in angle of attack would cause the airplane to stall and drop back onto the runway. In addition to being an unsafe operation, this practice would result in a run longer than necessary to ultimately get airborne. Therefore, some additional margin of speed is added to the stall speed to arrive at what is considered both a practical and safe takeoff speed. Usually, 20 percent above stall is considered an adequate margin of speed. Thus, if your stall speed is 60 kn, 72 kn would be considered takeoff speed. Remember that this stall speed is for the configuration (gear and flap setting) and weight at which you are operating.

The problem in takeoff performance is to determine the runway distance required. The actual distance for a takeoff is highly dependent on pilot technique. One pilot may use up a lot more runway for takeoff than another under the same conditions and with the same airplane. An equitable measure of takeoff distance, then, is that distance required to reach a specified value of takeoff speed. Whether one actually lifts off right at that distance and speed is immaterial. The important fact is that at this point it *could* be lifted off.

Takeoff performance is now reduced to the problem of determining the distance required to accelerate from zero velocity (at the beginning of the run) up to takeoff velocity. If we use the value of 1.2 V_S for takeoff velocity, the situation is further defined in terms of specific numbers.

Forces Acting

How fast the airplane accelerates depends on the quantity of the forces acting. Normally in flight we have drag tending to retard our forward progress and thrust to overcome this force. We also have these forces acting on the airplane during takeoff, but in addition we have a frictional force of the tires rolling on the runway. This force is somewhat alleviated, as the takeoff run progresses, by the lift reducing the weight on the wheels. This effect is rather slight, however, for many modern tricycle-gear airplanes, unless a definite effort is made to raise the nose slightly (and, hence, increase angle of attack) early in the run. Too much angle of attack, on the other hand, increases drag and defeats the purpose of reduced friction.

Another force acting to retard forward motion is the inertial force of the airplane. This force is explained by Newton's First Law of Motion, which says that "bodies at rest tend to remain at rest." All of these forces are shown in Fig. 5-1. Notice that drag, friction, and inertia all tend to retard the forward motion and, hence, acceleration. The only force acting for us is the thrust. Drag, friction, and inertial forces can all be lumped together and called "retarding force." A greater difference between thrust and retarding force will result in faster acceleration. Both frictional force and inertia are proportional to weight, so lower weight means lower retarding force, and hence, faster acceleration. Of course, faster acceleration means the ability to reach takeoff speed in a shorter distance.

Airports are ideally laid out on a perfectly flat surface. Since this is not always possible, runways usually have some amount of slope to them. If a fair amount of slope exists in a runway, then an additional gravity force is present during takeoff. An upslope run means a gravity force tending to retard motion, and a downslope run results in a gravity force which adds to the acceleration. Since standard takeoff

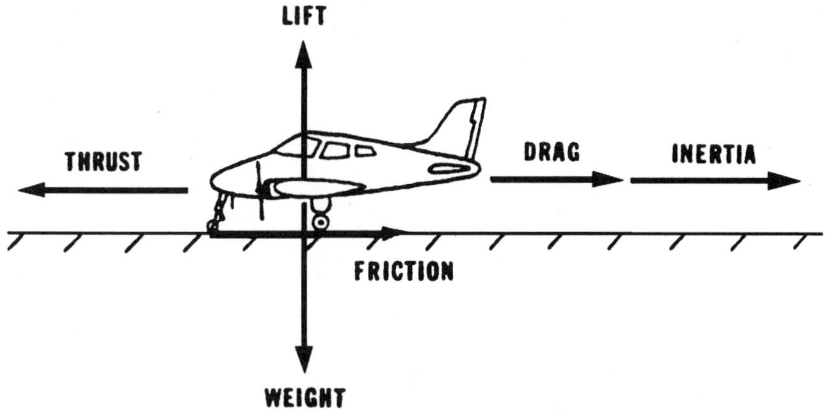

Fig. 5-1. Forces acting on the airplane during takeoff run.

data is desired for a no-slope condition, it is necessary to perform tests at an airport with a relatively level runway. Runway gradients (slopes) of 0.3 percent or greater are supposed to be indicated in the Airport/Facility Directory. However, not all airport listings include this information. If available, it is best to consult the engineering drawings for the airport construction. The airport manager should have such drawings. If you are operating from a very small field, such information may not exist. If you are uncertain of the slope of your strip or, if it has appreciable slope, it is wise to go to another airport where conditions are known to be better to run the takeoff tests.

Surface conditions of the runway determine the amount of retarding frictional force. Grass runways have more friction than hard-surfaced runways, for example. Thus, the data obtained will be for the type of surface that you use in the tests. If you want takeoff distance for a grass strip, you will have to run the tests on a grass strip. If you are based at such a strip and want takeoff performance for a hard-surfaced runway, then, again you will have to find such a facility at which to perform the tests.

Test Procedure

Since the takeoff problem has been explained as one involving the acceleration of the airplane, a *possible* takeoff distance can be determined by measuring how long a distance it takes to accelerate to takeoff velocity. Such a test of distance to accelerate to a certain speed could also be applied to an automobile or any other land vehicle. Thus it should be obvious that the test procedure does not involve a noting of the point of liftoff. In fact, it is not necessary to even take off at all in order to determine where the airplane *could* take off. All that is necessary is a takeoff run, keeping track of the time to cover various increments of distance during the run. Usually, the run will be of necessary length so that not enough stopping distance is left on the runway. For this reason, it is wise to go ahead and complete an actual takeoff. This procedure also assures that you have continued the run long enough to reach takeoff speed. The point is that with a sufficiently long runway, an actual takeoff does not have to be made to perform the test.

Prior to starting the test runs, some markers must be located along the runway at a spacing of about 100 to 200 ft. Existing runway lights serve well as distance markers. Their spacing can usually be found on the airport layout drawings. If not, the distance between them can be measured with a tape measure. If your runway has no lights, wooden stakes can be placed every 100 ft for the first 300 or 400 ft and, then every 200 ft up to about 1500 or 2000 ft. Be sure to get the airport owner's permission before doing this.

Load the airplane to approximately the standard weight value that you have decided upon. There are weight corrections that can be applied for tests at nonstandard weight, but it is best to operate at the weight for which data are desired.

To begin the test run, taxi up to the first marker. The pilot should very slowly ease up to this point until the spotter just has the first marker spotted as shown in Fig. 5-2. Spotting is done by sighting along a point on the airplane to a marker on the ground. The spotter should keep his head in a fixed position so that the angle of sighting does not vary. If the airplane is low-wing, the leading edge of the wing tip or tip tank is a good spotting point. For a high-wing airplane a point on the wing strut, a pitot tube, or some other protrusion below the wing can be used. The right front seat is usually the best position for the spotter with the markers laid out on the right side of the runway.

Fig. 5-2. Spotting the runway lights to be used as takeoff distance markers.

With the airplane lined up at the first marker, considered the "zero point," the pilot should then apply full power while holding the brakes. When the brakes are released, start the timing. This procedure does not necessarily shorten the takeoff run. It merely ensures that the run is started at a definite point in time. The airplane should then be allowed to accelerate in a normal manner down the runway. Don't attempt to lift the nose prematurely. If the airplane is a taildragger, raise the tail as soon as possible to the normal takeoff position. It is important to keep the airplane headed as straight as possible down the runway so that the wing passes each marker at the same angle.

As each marker is passed, record the time. There are several ways in which to do this timing. Probably the simplest way is to use a stopwatch which has a "lap" feature. This is one in which the reading can be stopped momentarily, but the actual timing continues from the initial starting point. This procedure allows the time at one marker to be read and recorded, and then the original time switched back on so that the procedure can be repeated at each successive marker.

An alternate procedure is to use a portable tape recorder instead of a stopwatch. Start the recorder prior to release of the airplane and call "start" or "go" into the recorder as the run begins. Then, as each marker is passed, call "one," "two," "three," and so on, into the recorder. Be sure to speak in a short, abrupt manner so that a distinct time of marker passing can be determined. Then, after the flight, the recording can be played back and the time to each marker determined with a stopwatch.

As a last resort, a single continuously running stopwatch can be employed. For this method one person has to continually monitor the stopwatch and catch the reading as the spotter calls out the marker. This works pretty well for the first few markers, but as the acceleration builds up, the last few are passed quite rapidly. You might try several methods, or original variations of them, to see which works best for you.

Photo Recording Method

Probably the best method for recording the passing of the distance markers is the use of a rapid-wind camera to photograph a continuously running stopwatch. Automatic winding cameras are readily available nowadays, even among the popular "point and shoot" variety. The close focusing lens required, however, may not be available on such cameras, making the single lens reflex the best choice for this purpose. The procedure here is to rig up a fixture to hold both the camera and the stopwatch, such as shown in Fig. 5-3. Make sure it is long enough to allow the lens to focus on the watch. A macro lens is an excellent choice.

Select a stopwatch that indicates in hundredths of a second. The camera will then record the time to that accuracy provided the shutter speed is set higher than 1/100 second. About 1/200 is recommended, but 1/125 is sufficient if the light in the cockpit requires it. Higher speed film will help in this regard, probably 400 ISO. Color film is acceptable, either for producing prints or transparencies (slide film), but black and white film is also quite sufficient. If you are a photo buff, and can process your own film, or know someone who does, this can be the simplest approach. You don't even need to print the film, but simply develop it, and read the stopwatch readings from the negatives. Figure 5-4 shows a typical shot.

Fig. 5-3. Mounting device for camera and stopwatch to record time passing distance markers.

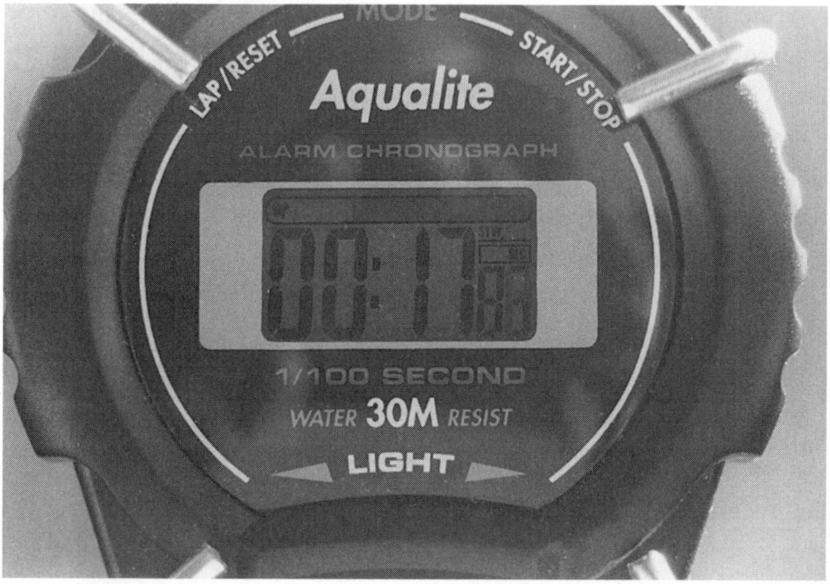

Fig. 5-4. Shot of stopwatch showing time for passage of marker taken during actual takeoff test.

Prefocus the camera to sharply record the stopwatch. If using an autofocus camera, select manual focus mode. This prevents the delay in the shutter releasing while the focusing is accomplished. The shutter speed is critical, so don't use a program mode, which may select a very low speed. Shutter-priority is acceptable, but a full manual metering mode is best. Even if the light is too low for an accurate exposure setting, you can probably get away with a bit of underexposure. You are not trying to produce a masterpiece, but only a readable image.

At the release of the brakes, hit the start button on the stopwatch. Then sight the markers just as with any other recording method. *Don't look through the camera viewfinder.* Look out at the runway. At each marker, push the shutter release. Try not to miss any markers, since you want to have them in order. After liftoff, you can stop the watch, and reset it for another run. Skipping a frame is a good way to mark the place on the film between runs. You can do this by placing the cap over the lens and shooting one frame. (Don't forget to remove the cap before the next run.)

If you develop your own film, or if the film lab will do it, keep the roll uncut. This way, all your shots will be in order. Also make some written notes on the conditions for each run. That is, the weight, wind, flap setting, pressure altitude, and temperature. You could also note whether it looks like a good run or not, meaning whether you started right at the release point, and/or hit the shutter pretty much right at the markers or not. Another way of more positively identifying the specifics of a run is to mark them on a small card with a fairly broad tipped pen, place it over the watch face, and shoot one frame to record the data.

Velocity Calculation

Regardless of the method used, the time to pass each marker must be obtained, and as accurately as possible. This time is then plotted for the respective distances of the markers as shown in Fig. 5-5. The average speed between each pair of markers is then determined by calculation. Since average speed is, simply, distance divided by time, take

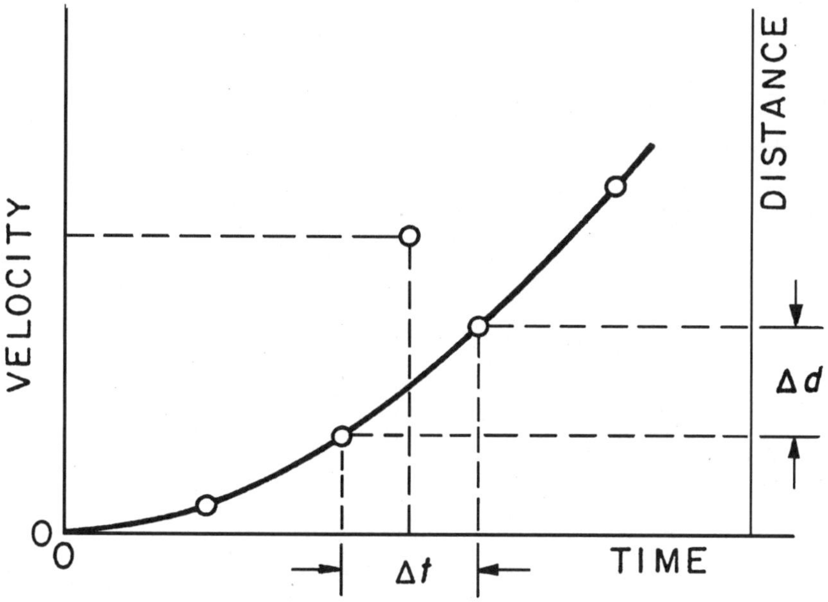

Fig. 5-5. Plotting distance and average velocity during takeoff run.

the distance between two markers, as shown, and divide by the difference in time between these same markers. Establish a velocity scale on the left hand side of the chart and plot the average velocity at a point halfway between the marker positions. The equation for velocity is:

$$V_{average} = \frac{\Delta d}{\Delta t}$$

If the distance is measured in feet and the time in seconds, then the velocity will be in feet per second. To convert to mph, this value must be divided by 1.467. Thus:

$$V_{av}\ \text{mph} = \frac{\Delta d}{\Delta t(1.467)}$$

To convert to knots divide by 1.687 instead of 1.467. After computing the average velocities between each set of markers and plotting them, draw a smooth curve, fitted as closely as possible to both the distance points and the velocity points.

If the velocity points calculated by this method do not fall on a smooth curve, a better procedure is outlined as follows: Plot the points for the time to various distance markers, as in the above procedure, and draw a smooth curve through these points. Then, take even increments of time, for example, every two seconds and find the distances corresponding to these times from the curve drawn. Now, use the difference in these distances and the Δd in the above equation and the Δt that you have selected (such as the two seconds mentioned) to calculate average velocities. The smaller the time increment chosen, the more accurate the velocity curve will be.

Now determine your possible takeoff speed as 1.2 times the stall speed. This value must be TAS, so take the indicated stall speed for the configuration being flown (gear down, we hope), correct it first to CAS by your calibration curve, and then correct this for density to TAS. Note that this procedure requires the stall speed to first be determined by methods in the previous chapter and an airspeed calibration to be done as outlined in Chap. 3.

Once the takeoff speed in TAS is determined, enter the chart you have constructed at this value on the velocity scale (left hand side). Move right until you hit the velocity curve. Then go vertically up or down as necessary to hit the distance scale. Now read the distance scale directly to the right of this point. This is your takeoff distance. The procedure is shown in Fig. 5-6.

Wind Correction

Tests should be run in wind as near to dead calm as possible. Light headwinds of 10 kn or less can be tolerated and corrected for as long as they are steady. Some means of measuring the wind must be employed. There are a number of handheld anemometers that can be used for this purpose. Be sure to measure only the headwind component of the wind by pointing the anemometer straight down the runway (Fig. 5-7).

If a handheld anemometer is not available, it is possible to make use of the wind information as reported by the tower, FSS, or whatever facility is available. Winds as reported by such facilities will be given in the true direction from which they are blowing. It will be necessary to compute the headwind component of this wind if it is not

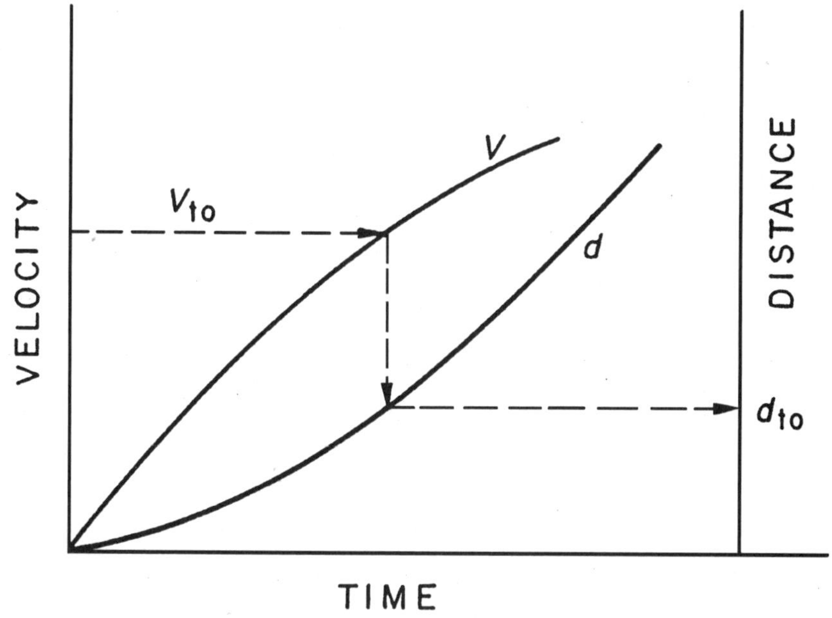

Fig. 5-6. Finding takeoff distance from plot of velocity and distance versus time.

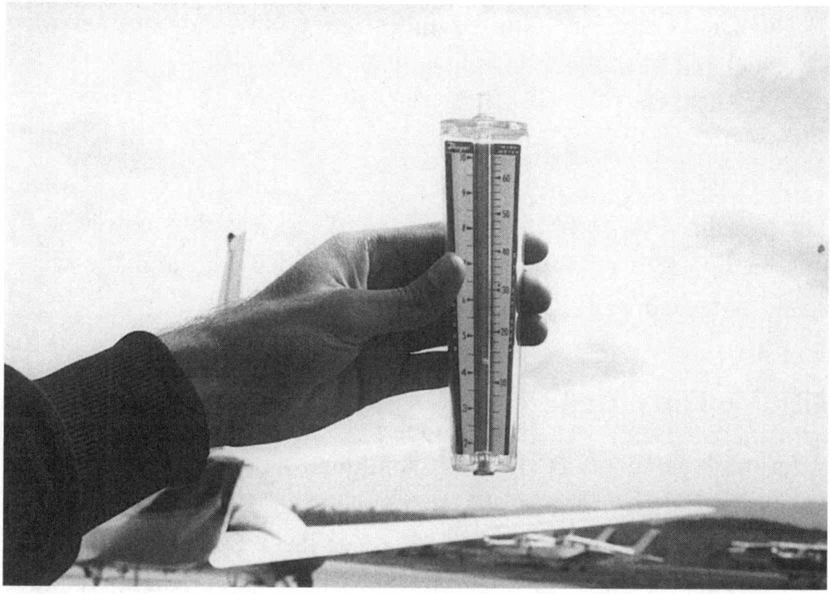

Fig. 5-7. Measuring headwind component with hand-held anemometer.

very nearly directly down the runway. Headwind components can be calculated by use of a wind component chart shown in Fig. 5-8. Before using this chart, you must first determine the angle between the wind and the runway. For a wind blowing from 270°, for example, and a runway numbered 24 (approximately 240°), the angle between the wind and the runway is 270° minus 240° or 30°. Suppose a wind were blowing at this angle at 15 kn. From the chart, going out to 15 kn on the 30° line and then reading directly to the left, you would obtain a headwind component of 13 kn. Remember that wind is given in true direction and runway headings in magnetic. If much variation exists at your airport location it will have to be taken into account. In central Pennsylvania, for example, the variation is 8° W. A true direction of 270° would, therefore, be 278° magnetic. In New England and the northwestern part of the United States, the variation can be 20° or more. Runway headings, of course, are rounded off to the nearest ten degrees. If you didn't know the exact direction of layout, just adding a zero to the number could result in an error of no more than plus or minus five degrees. The runway 24 used in the above example could actually head anywhere from 235° to 245°. This amount of error is negligible in calculating headwind components of wind.

The takeoff distance required under zero wind conditions can then be calculated from the following equation:

$$d_{0\text{wind}} = \left(\frac{V_{\text{to}}}{V_{\text{to}} - V_{\text{wind}}}\right)^2 d_{\text{wind}}$$

The measured takeoff distance with wind, d_{wind}, is thus corrected by this factor using your calculated takeoff velocity V_{to} and the wind-speed as measured V_{wind}.

Altitude Correction

If the tests are conducted under conditions other than standard sea level (as they probably will be) a correction must be applied to yield the distance that would be required at sea level. Such corrections are not as exact as some relationships and various approximations have been devised. If the test condition is not too different from standard

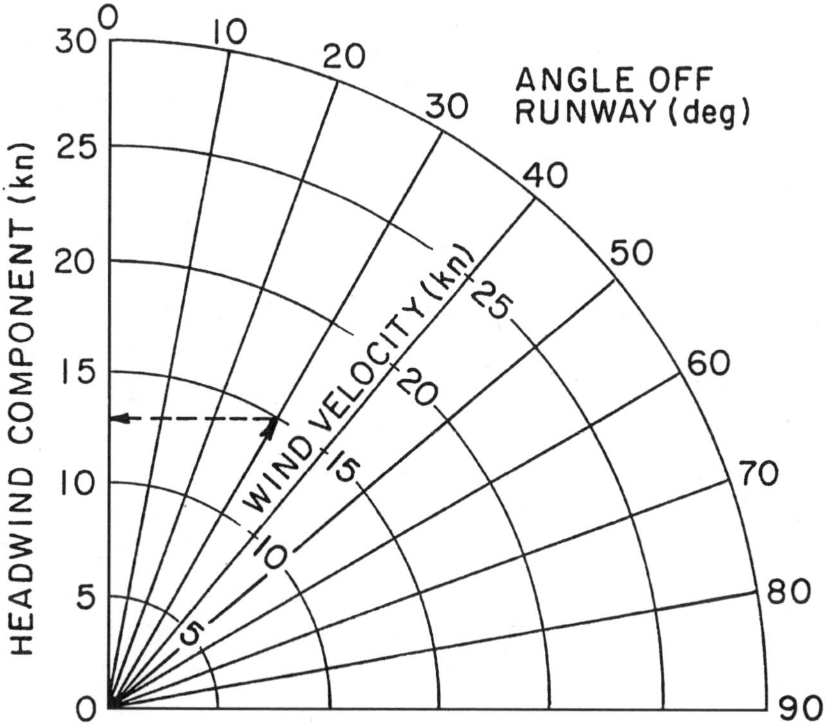

Fig. 5-8. Chart for determining headwind component.

(that is, at a fairly low density altitude), a simple correction is possible using the density ratio. First find your actual density altitude from Fig. 1-3 or A-1 and then the density ratio for this altitude from Fig. 1-4. The standard sea level takeoff distance is determined by multiplying the test distance by the square of the density ratio.

$$d_{\text{std}} = d_{\text{test}} \left(\frac{\rho}{\rho_0} \right)^2$$

The standard sea level, zero wind, value of takeoff distance can then be adjusted to other conditions by use of approximate correction factors. For airplanes with fixed-pitch propellers, the distance at any altitude (including fairly high-density altitudes) can be approximated by dividing the standard sea level distance by the square and the square root of the density ratio for the desired altitude.

$$d_{\text{alt}} = \frac{d_{\text{std}}}{(\rho/\rho_0)^2 \sqrt{\rho/\rho_0}} \quad (\rho/\rho_0 \text{ for desired altitude})$$

This relation does not apply too well to constant-speed propellers. If you are testing an airplane so equipped, another method of approximating takeoff at altitude is by use of a Denalt computer. This is a device developed by the FAA to determine performance at various altitudes. For takeoff distance you set the appropriate air temperature and read opposite the desired pressure altitude a correction factor. Then simply multiply the sea level takeoff value by this factor to get the distance at altitude. There is one of these devices for constant-speed propellers and another one for fixed-pitch propellers. The Denalt computer for fixed-pitch propellers could also be used instead of the method given above. Figure 5-9 shows a Denalt computer.

It should be noted that these methods for obtaining takeoff performance at altitude by calculation from the sea level are only *approximations*. Experience has shown that the equation involving the density ratio sometimes yields results a bit short of the actual

Fig. 5-9. The Denalt Performance Computer used to determine takeoff distance at altitude.

values as tested. On the other hand, the Denalt computer is a bit conservative and usually gives a little higher than actual. If you really want to know the takeoff distance at a fairly high density altitude (5000 ft or more), it would be wise to run actual tests at those altitudes. The procedure would be the same as the test outlined previously, except that it would be repeated at various density altitudes. Such a procedure would mean finding higher altitude airports or testing on higher density altitude days (usually meaning hotter days) or a combination of both.

The takeoff distance for any other weight can also be calculated by multiplying the standard distance by the square of the new weight desired divided by the square of the standard weight.

$$d_{\text{new wt}} = d_{\text{std}}\left(\frac{W_{\text{new}}}{W_{\text{std}}}\right)^2$$

Starting with your standard sea level, zero wind, standard weight value, the takeoff distances for a number of other altitudes and several weights should be calculated. These values can then be plotted to give a workable takeoff performance chart as shown in Fig. 5-10. Note that these curves are almost straight lines but not quite.

Takeoff distance should also be corrected for runway slope, as previously mentioned. This procedure, however, is a bit more complex, and involves some trigonometry. It also requires some detailed information on the slope of the runway being used for the tests, including the variation of slope over the length of the runway, which may not be readily available. Since this correction is somewhat complicated, it will not be addressed here. Nevertheless, it should be recognized that runway slope can have a significant effect on takeoff performance, in some cases, even more so than wind. Therefore, try to conduct the tests on a nearly level runway.

Takeoff Distance to Clear Obstructions

Up to this point we have been considering takeoff distance as only that distance required to break ground. Another important consideration in takeoff is the ability to climb over obstacles that may exist near

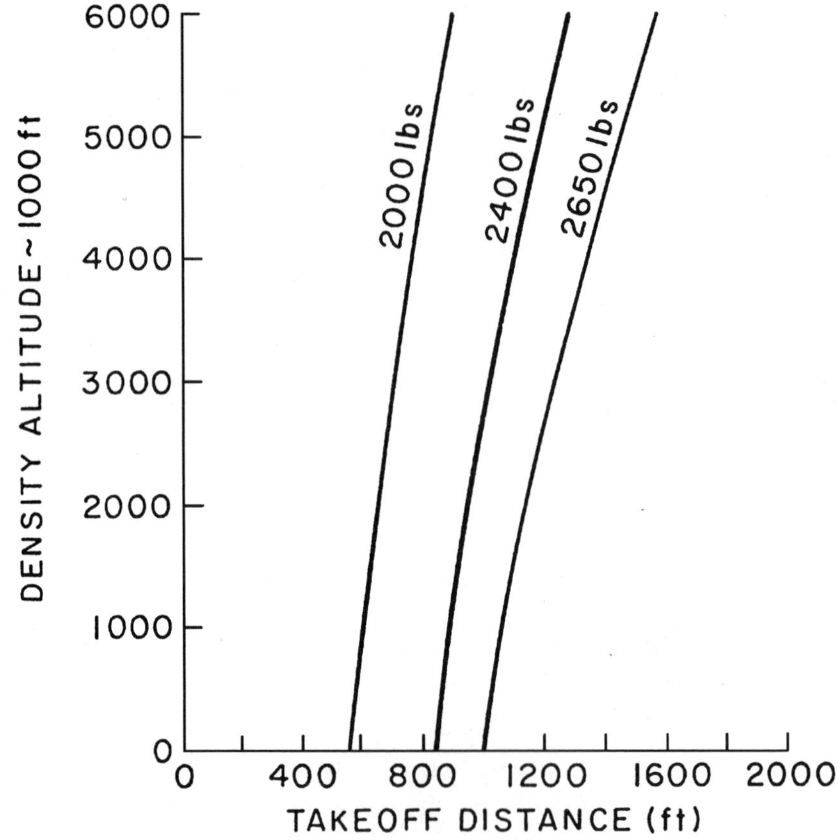

Fig. 5-10. Typical plot of takeoff distance for various gross weights at various density altitudes.

the end of the runway. Federal Aviation Regulations specify 50 ft as the standard height of an obstacle that must be cleared. In addition to ground run distance, therefore, you will usually find in airplane manuals the distance to takeoff and clear a 50-ft obstacle.

In determining total distance to clear a 50-ft obstacle, we usually break the takeoff distance into two parts (Fig. 5-11). The first is the ground run distance, which we have previously been discussing. The second is the air distance. Air distance is the horizontal distance from the lift-off point to the point where the altitude is 50 ft. In other words, air distance is the portion of the total takeoff run during which the airplane is in the air.

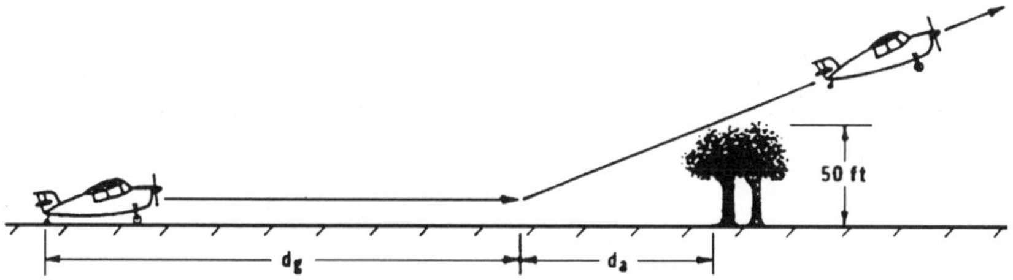

Fig. 5-11. Takeoff over a 50-ft obstacle showing ground distance and air distance.

Measuring air distance, experimentally, is a bit complicated. For best results a vertically oriented radar is used. A movie camera photographing through a measured grid can also be used. Since most of us do not have such elaborate equipment, accurate location of the 50-ft altitude and the horizontal point at which it occurs is not easy. Even if it could be determined, the performance in this area is so dependent on pilot technique that many varying distances would result. A good determination depends on the pilot rotating at the correct speed to break ground and then attaining and holding the proper climb speed. FAR Part 23 states that at the 50-ft altitude the airplane must have reached 1.3 V_s or best angle-of-climb speed (V_x) plus 4 kn. All of these requirements impose considerable demands upon the pilot, especially when you consider that all of this has to be done within the first 50 ft of altitude. Therefore, even accurate measurements would not yield very reliable information unless quite a number of takeoffs were recorded and an average value determined.

A simpler method of determining air distance is to calculate it rather than measure it from tests. The calculated air distance can then be added to the ground distance which was determined by tests. This can be done rather easily for airplanes with rather shallow climb angles, typical of most light planes. The assumption made is that the airplane lifts off at its best rate of climb speed. Rate of climb and best rate-of-climb speed will be determined in the next chapter. It will be seen that this best rate-of-climb speed (V_y) is very close to V_x plus 4 kn and also very close to our previously specified lift-off speed of 1.2 V_s. At shallow climb angles, you can assume that the horizontal speed of the airplane is approximately equal to its forward speed. Remember that in a climb, the forward speed is in a direction somewhat inclined to the horizontal. With this assumption the air distance to reach a certain altitude can be calculated simply as the ratio of the horizontal to vertical speed times the altitude:

$$d_a = \frac{V}{RC}\ (50)$$

The velocity used can be takeoff velocity and the rate-of-climb value obtained from the tests in the next chapter. Use the rate of climb for the appropriate altitude. Since velocity is usually measured in knots or miles per hour and rate of climb in feet per minute, some conversion factors must be included to make a workable equation. The results are as follows:

$$d_a = (4401)\ \frac{V_{to}}{RC}\ \text{(for } V \text{ in mph)}$$

$$d_a = (5061)\ \frac{V_{to}}{RC}\ \text{(for } V \text{ in kn)}$$

The total distance, then, to clear a 50 foot obstacle is the ground distance as previously determined plus the air distance:

$$d_{50\text{ ft}} = d_g + d_a$$

Charts could also be made to include air distance and plotted similar to the ground-run distance as in Fig. 5-10.

Takeoff Test Example

Test Conditions:

- Gross Weight: 2200 lbs.
- Pressure Altitude: 2000 ft.
- Temperature: 38° F.
- Headwind (measured): 4 kn (5 mph).

• Distance versus time as listed:

D(FT)	T(SECONDS)
100	5.5
200	8.0
300	10.0
400	11.5
600	14.0
800	16.5
1000	18.5

1. Plot distance versus time as in Fig. 5-12, and draw a smooth curve through the points.
2. Take time increments of 2 seconds and divide into corresponding increments of distance from the curve to calculate average velocities. Divide by 1.467 to get velocity in mph. Example:

$$d_{\text{av}(4\text{–}6\text{ sec})} = \left[\frac{108' - 45'}{6\text{ sec} - 4\text{ sec}}\right]\left[\frac{1}{1.467}\right] = 21.5\text{ mph}$$

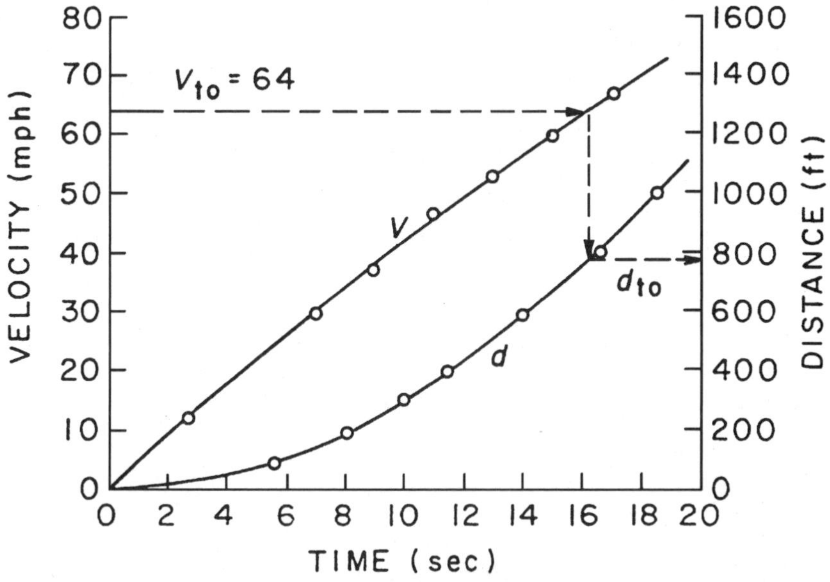

Fig. 5-12. Finding takeoff distance for the example given. For a takeoff velocity of 64 mph, the takeoff distance is 780 ft.

3. This point is plotted for the velocity at 5 seconds, the average between 4 and 5. Other average velocities are calculated similarly and plotted as the velocity curve in Fig. 5-10.

4. The stall speed for this configuration is established (by prior test) at 53-mph TAS. Takeoff velocity is calculated as 1.2 V_S.

$$V_{to} = 1.2\ (53) = 63.6 \text{ mph}$$

5. This figure is rounded off to 64. Enter the chart plotted at this velocity, go straight across until the V curve is reached. At this point go vertically down to the d curve. Directly to the right of this point read 780 ft, the *test* takeoff distance.

6. This number is then corrected to zero wind from the test wind condition:

$$d_{0\text{wind}} = \left[\frac{V_{to}}{V_{to} - V_w}\right]^2 d_{test} = \left[\frac{64}{64 - 5}\right]^2 780 =$$

$$1.177\ (780) = 918 \text{ ft}$$

7. This figure is further corrected to standard sea level. From pressure altitude of 2000 ft and T = 38°F, by use of Fig. 1-3, density altitude is 1000 ft. From Fig. 1-4, the density ratio is 0.97:

$$d_{std} = \left[\frac{\rho}{\rho^0}\right]^2 d_{test} = (0.97)^2\ (918) = 864 \text{ ft}$$

Takeoff distance for 2200 lb, zero wind, and standard sea level is, thus, established as 864 ft.

8. Takeoff for other gross weights at sea level are calculated by:

$$d_{\text{new weight}} = 864\left[\frac{W_{new}}{2200}\right]^2$$

Example:

$$d_{2000} = 864\left[\frac{2000}{2200}\right]^2 = 714 \text{ ft}$$

9. The distances for any of these weights at altitude are calculated by:

$$d_{alt} = \frac{d_{SL}}{(\rho/\rho_0)^2{}_{alt}\sqrt{\rho/\rho_0}}$$

Example:

$$d_{(3000\text{ ft, }2000\text{ lbs})} = \frac{714}{(0.915)^2\sqrt{915}} = 892\text{ ft}$$

CHAPTER 6

Climb Performance

The unique characteristic of an aircraft, compared to land-based vehicles, is that it can move in a third dimension. That dimension is vertical to the surface of the earth. The performance of the airplane in that dimension is, therefore, an important mark of the achievement of its intended purpose. Vertical motion away from the surface is referred to as *climb.* How well the airplane performs in this direction is usually determined by its rate of climb. The rate of climb depends both on the power output of the airplane's engine and the drag of the airplane. In order to understand the behavior of the airplane in climbing or descending flight, it is first necessary to discuss both drag and power.

Drag

Aerodynamic drag is the resistance of the air which tends to retard the airplane in forward flight. Since forward flight is necessary to create lift, drag consequently plays a large role in climb performance.

Total drag depends on the configuration of the various airplane components and the way in which they are combined. The various ways in which the air creates drag on a component leads to several definitions for different kinds of drag. For a low-speed airplane (below Mach .7) there are two major types of drag—parasite and induced.

Parasite drag is the resistance of any body to the air through which it is moving (or which is moving over it). It is generally broken down further into two parts: *form* drag (sometimes called *pressure* drag) and *skin friction* drag (sometimes called *viscous* drag).

The plate set perpendicular to the airstream as in Fig. 6-1*a* encounters practically all form drag, while the plate aligned with the airstream as in Fig. 6-1*b* encounters practically all skin friction drag. Since most bodies have a shape somewhere between these two extremes, such as the aerodynamic shape in Fig. 6-1*c*, they usually are subject to both types of drag. Thus, the term "parasite drag" is commonly used to denote the sum of both form and skin friction drag on a body.

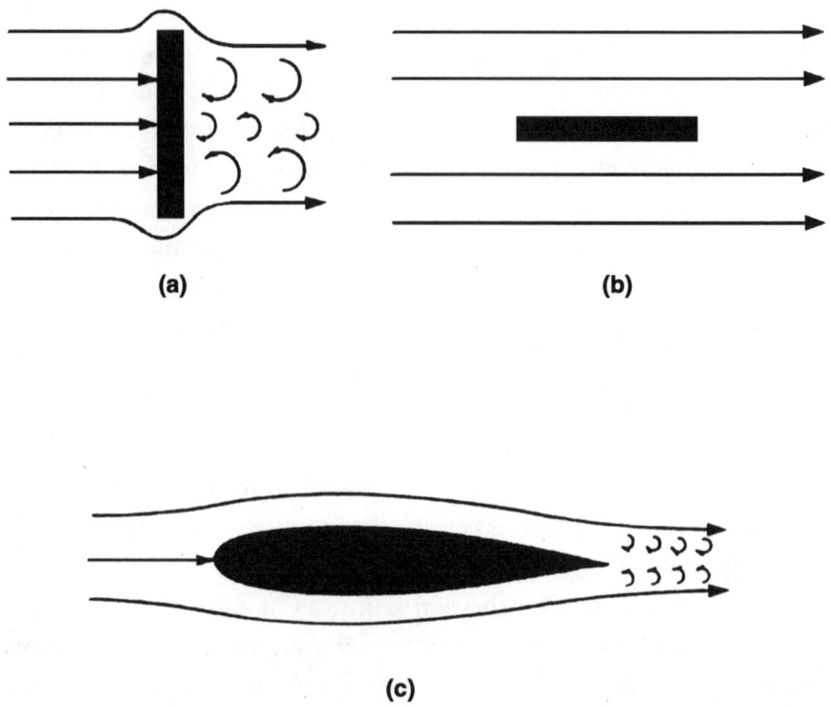

Fig. 6-1. Form and skin friction drag. (*a*) Thin plate in this position encounters all form drag. (*b*) Plate in this position encounters all skin friction drag. (*c*) Normal aerodynamic body has combination of form and skin friction drag.

Induced drag is an entirely different type of drag and applies only to the wing or other surface that creates lift. Lift is *defined* as the component of the aerodynamic force on the wing that is perpendicular to the velocity vector of the wind creating this force, such as in Fig. 6-2*a*. Because of the downwash, the wind vector is actually turned downward through a slight angle on an actual wing. The actual lift vector is therefore really perpendicular to this new velocity vector that results at the wing (Fig. 6-2*b*). In terms of our original definition of lift, that is, perpendicular to the velocity coming toward the airplane (or opposite to the flight path of the airplane), the lift is only the vertical component of the actual vector. Breaking up the lift in this manner also gives a component in the direction of the original velocity vector. Since forces in this direction are defined as drag forces, this component of the actual lift vector is termed "induced drag." It is an unavoidable consequence of the creation of lift by a wing.

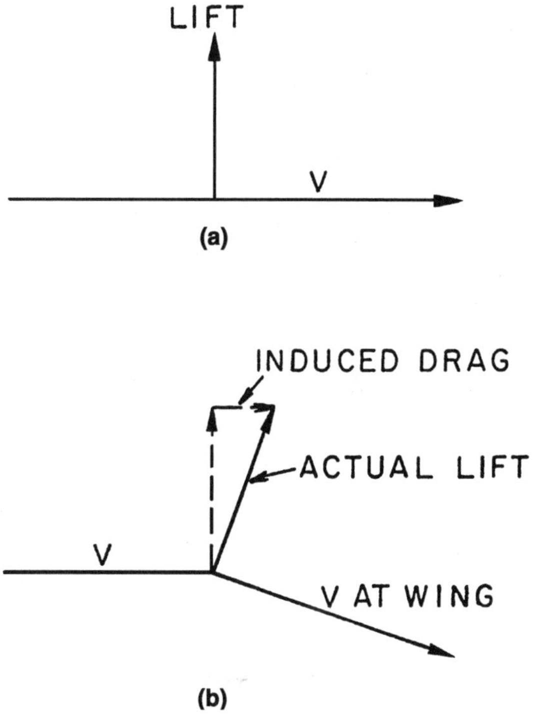

Fig. 6-2. Induced drag. (*a*) Lift is normally considered as force perpendicular to the airstream. (*b*) With downwash, the airstream is tilted downward at the wing, tilting the lift vector backward, producing a horizontal component of lift known as induced drag.

Since parasite drag is caused by the dynamic pressure of the air, and dynamic pressure is proportional to the square of velocity (dynamic pressure = $\frac{1}{2}\rho V^2$), it follows that parasite drag is also proportional to the square of velocity. Induced drag, on the other hand, is proportional to the downwash and this increases at high angles of attack associated with low speeds. It actually works out that induced drag is *inversely* proportional to the square of velocity. In Fig. 6-3, the plot of drag versus velocity shows these opposite trends for parasite and induced drag. The total drag is the sum of parasite and induced and results in the characteristic curve as shown. This curve represents a plot of the drag equations:

$$D = D_{par} + D_{ind} = K_1 V^2 + \frac{K_2}{V^2}$$

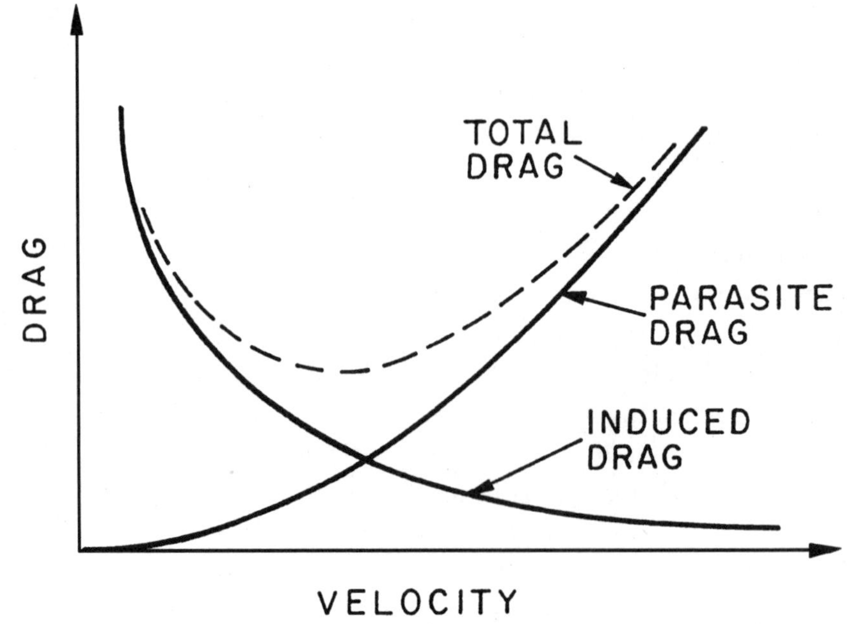

Fig. 6-3. Relationship of parasite drag, induced drag, and total drag with airspeed.

Notice that total drag increases from some minimum point with *both* increase and decrease in velocity. This fact is very important to airplane performance, particularly to climb performance.

Power Required

Drag, then, represents the force that must be overcome by the thrust of the powerplant. Since the power of a reciprocating engine is easier to determine than the thrust of a propeller-engine combination, drag is often defined in terms of *power required*. Power is equal to a force times a velocity. Thus, if the drag force in the form of the above equation is multiplied by another velocity term, the power-required equation ends up as:

$$P_{\text{required}} = K_1 V^3 + \frac{K_2}{V}$$

The first term is called the parasite power term and is proportional to the cube of velocity, while the second term represents the induced power and is inversely proportional to the first power of velocity. Power is usually expressed in units of horsepower and results in the same equation except that the constants K_1 and K_2 have different values. When plotted, this equation yields a typical curve as in Fig. 6-4, commonly known among aviation in-groups simply as the *power curve*. This curve represents the power that is required at any velocity to maintain steady, level flight. For example, at 150 mph this particular graph shows that 120 hp is required.

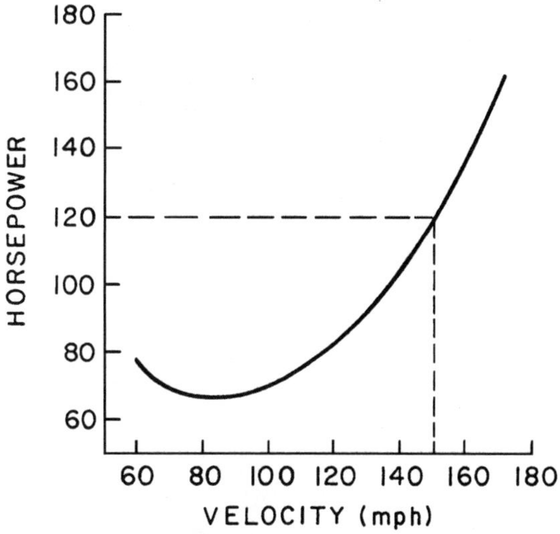

Fig. 6-4. Horsepower required versus velocity.

Power Available

Whether the airplane can actually fly at a certain value of power required (such as the 120 hp mentioned above) depends on whether that power is available from the powerplant. Obviously, if the engine is rated at 100 hp then it could never develop 120 hp. This would mean that the airplane could never fly straight and level at 150 mph in the example given.

The amount of power that can be obtained from the powerplant is termed *power available*. For any given engine, this power can be varied with throttle setting (and propeller pitch, if constant-speed). Also, the power at a given setting will decrease with altitude. Reciprocating engines that are normally aspirated (not turbocharged) display a relationship with altitude as shown in Fig. 6-5. This chart shows the horsepower at any altitude for a given rpm setting. The dotted line shows that at 5000 ft and 2000 rpm the engine delivers 100 hp. This power is really the power available right at the crankshaft of the engine and is referred to as *brake horsepower* or bhp. If the engine is geared down to a lower speed at the propeller shaft, the power available there is called *shaft horsepower* or shp. The shaft horsepower is a little less than brake horsepower if gearing is employed due to frictional losses. For direct drive engines, however, shaft horsepower is the same as brake horsepower. Note that as altitude is increased at any

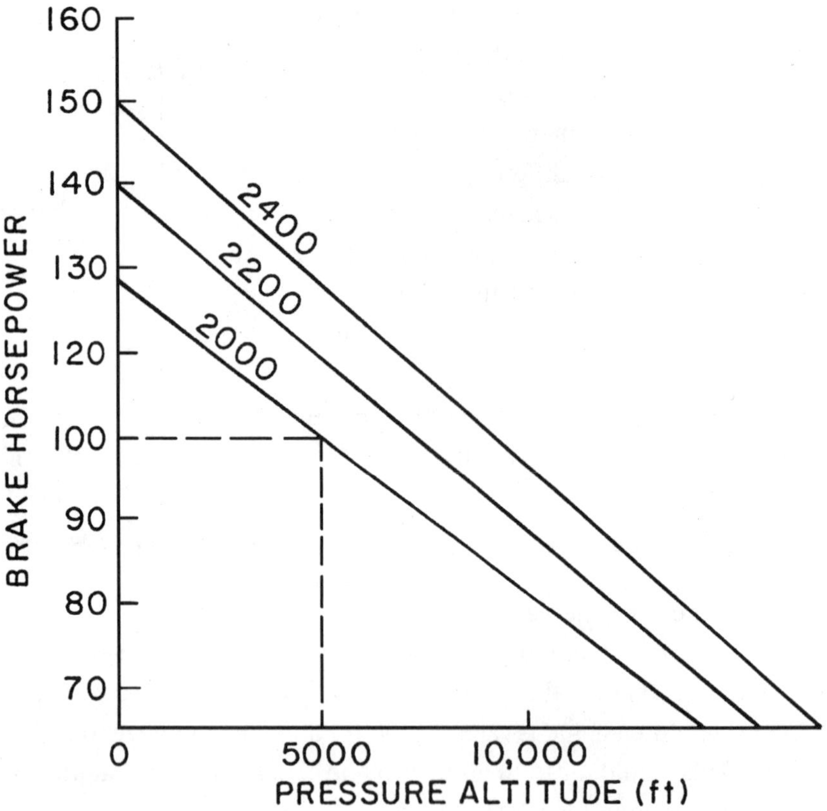

Fig. 6-5. Engine power chart showing variation with altitude.

specific value of revolutions per minute, the power drops off. This is because the density of the air gets less with altitude and the engine power output is proportional to the air density.

Regardless of the altitude, however, all of the brake horsepower (or shaft horsepower for geared engines) does not get converted into thrust. Thrust is provided by the propeller and a certain amount of power is lost in converting the torque of the engine into thrust force. The amount of power that does finally get converted into thrust is referred to as *thrust horsepower* or thp. It is this thrust horsepower that must be counted on to match, or exceed, the power required due to drag.

The propeller acts basically like a wing in creating lift. Its "lift" vector however is pointed generally forward and the "lift" force that it creates thus becomes thrust. Since the propeller has an airfoil section and works like a wing, it also creates a drag. Hence, some power is lost to the drag of the propeller. It works most efficiently at its maximum L/D (lift to drag) ratio. The propeller, however, is moving through the air at a forward velocity equal to that of the airplane, and also at some rotational velocity. A section of blade is shown in Fig. 6-6. Various combinations of forward and rotational speed give different L/D ratios to the propeller and, consequently, different values of efficiency. A typical propeller efficiency chart is shown in Fig. 6-7. This chart shows that up to a point, higher forward speed at a fixed revolutions per minute value will increase propeller efficiency. If the propeller pitch can be varied, as with constant-speed propellers, the peak efficiency can be maintained over a longer range of forward speeds.

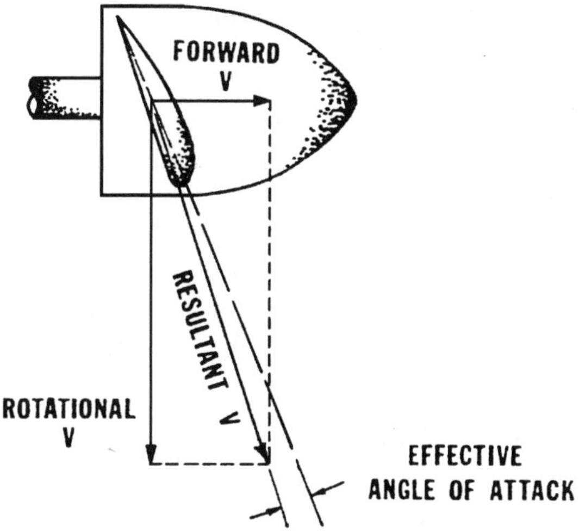

Fig. 6-6. Propeller motion showing forward velocity, rotational velocity, and resultant velocity from which angle of attack is determined.

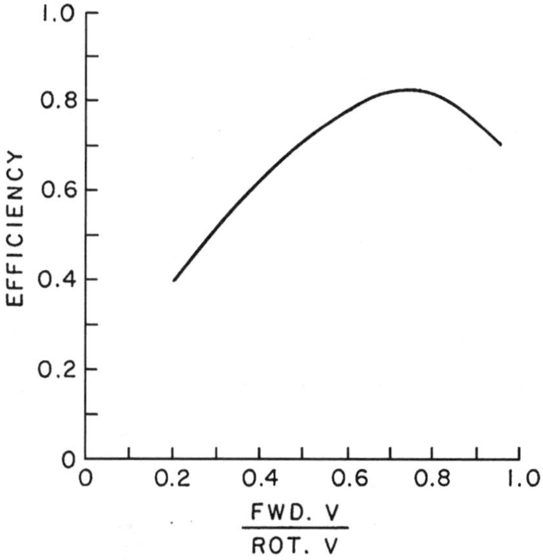

Fig. 6-7. Typical propeller efficiency curve as a function of forward to rotational speed ratio.

The horsepower available for thrust, or thrust horsepower, is, then, the propeller efficiency multiplied by the brake horsepower delivered by the engine. Since the brake horsepower is a fixed value at a certain altitude and power setting, the thrust horsepower varies with velocity proportionally to the way the propeller efficiency varies. Therefore, thrust horsepower available when plotted against forward velocity has the same shape as the curve in Fig. 6-7. Remember that this curve represents power *available*. Such power is variable anywhere from zero up to the maximum capacity of the engine depending on throttle setting. Usually, therefore, we are interested in the *maximum* power available. This would be the curve of propeller efficiency multiplied by the maximum brake horsepower for the particular altitude.

Rate of Climb

The airplane's ability to climb depends on the excess power available over that required. If the maximum power available is plotted on the same chart as the power required, the excess power available can easily be seen. Figure 6-8 shows such a plot. The excess power is the difference between maximum power available and power required at any given speed. The rate of climb is directly proportional to the excess power. Notice that in the figure, the *maximum* excess power is shown. This occurs at the velocity for rest rate of climb, usually designated as V_y. At speeds both above and below this velocity, the rate of climb will decrease. Where the curves cross, the excess power becomes zero and

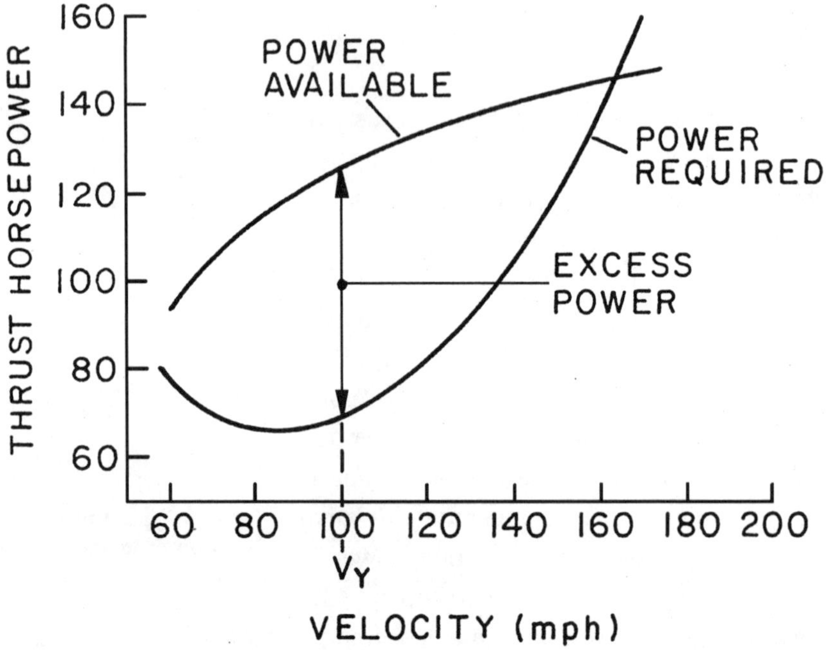

Fig. 6-8. Thrust horsepower versus velocity showing excess power as difference between power available and required.

climb is not possible beyond that speed. It is important, then, to know both the maximum rate of climb and the speed at which this rate occurs.

The curves shown are for only one altitude. At higher altitudes the maximum power available will decrease and the power required curve will be altered because of changes in drag. Therefore, the excess power becomes less with altitude and, hence, rate of climb becomes less. When there is no excess power at any speed, no further climb is possible and the airplane is said to have reached its *ceiling*. The altitude for absolute zero climb ability is called *absolute ceiling*. Another definition of ceiling is the *service ceiling*, which is a point more practical to consider. This is the altitude where the climb ability is reduced to 100 ft per minute. One could conceivably fly at this altitude but not quite at the absolute value.

Test Procedure

In measuring rate of climb by flight tests you might be tempted to simply read the rate of climb (or vertical speed) indicator. However, this instrument is not always that accurate and usually has a great deal of lag associated with it. A timed measurement of definite altitude change is much better.

Speed or velocity is really the rate of time in which distance is covered. Average velocity is very easily determined by dividing a certain amount of distance by the time it takes to go that distance. Rate of climb is also velocity, but in a vertical direction. The average rate of climb can, therefore, be determined by climbing through a certain amount of altitude and timing the climb with a stopwatch. The rate of climb is, then, the altitude in feet divided by the time in minutes. The value will be for a certain airspeed which must be held during the measurement.

Before beginning the flight tests, make sure that the airplane is loaded to the standard weight (maximum gross or normal operating weight). If a standard weight below max gross is chosen, it would be wise to load to somewhat above standard weight so that the fuel burnoff will reduce the weight to standard at about midway in the tests. In this way you will be operating closest to an average weight.

Weight is very important to climb performance. You may even want to repeat the entire procedure for various average gross weights, since the rate of climb will be significantly greater at lower weights. On the other hand, the highest weight gives the minimum climb performance. Thus, if max gross or near max is tested, you can be assured that at lower weights it would not be any less.

Select a pressure altitude and fly to that altitude. If terrain permits you may want to start with an altitude of 2000 or 3000 ft. Begin the tests at a speed about 5 kn above stall or high enough so that buffeting or other indication of imminent stall is not encountered. Descend to 300 or 400 ft below the test altitude and establish a climb by applying full power and holding a constant airspeed. At 200 ft below the specified altitude start timing. Climb at a steady velocity until 200 ft above the test altitude and stop the timing at this point. Then descend again and repeat the procedure through the same altitudes at other airspeeds. Five or 10 kn (or mph) increments are sufficient. The times should decrease to a point and then start increasing. Once this trend is definitely established, the peak rate of climb is past and no further speeds at that altitude need to be tested.

It should be obvious by now that very calm air is required for these tests, but this point cannot be emphasized too much. Vertical currents can exist even if the air appears to be smooth, with no turbulence. During the climb run, watch the vertical speed indicator. You are not going to rely on its reading, but is a good indication of a stable climb. If the VSI reading varies appreciably during the climb, there is probably vertical movement in the air, and that run should be discounted. It is a good idea to make several runs at the same altitude and airspeed. If the timing is consistent, the climb is probably fairly stable, and the test data reliable.

A correction is sometimes needed in the measured altitude change. Since this is done by altimeter reading, and the altimeter is based on standard lapse rate, the altimeter can have some error if the temperature is significantly different from standard. A simple correction can be made here. Take the actual temperature from the OAT (in °F), and then (in reducing the data) read the standard temperature for that altitude from a chart or table such as Fig. 1-3 or

Table 1-1. The actual altitude increment, Δh_{act}, can then be calculated from the measured increment, Δh_{meas}, by the equation,

$$\Delta h_{act} = \Delta h_{meas} \left[\frac{T_{act} + 460}{T_{std} + 460} \right]$$

Try this with one data point at each altitude. If there is not much difference, you can probably ignore the correction. For example, if your measured altitude change is 400 ft (a common figure used), and the corrected value is 405 ft, that is probably not significant, and within the possible error range of your measurements. If, on the other hand, you get 450 ft, it might be prudent to correct the other data points before proceeding. Note, that in the test example below, the temperature varies by only about 1 degree from standard at the test altitude, making the altitude change correction negligible. It was, therefore, ignored in the example calculation.

Determine rate of climb by dividing altitude change in feet by time in minutes for each speed tested. If time is measured in seconds, then multiply by 60 to get rate of climb in feet per minute.

$$\text{Rate of climb} = \frac{(\text{final altitude} - \text{initial altitude})}{\text{seconds}} (60)$$

Then plot the resulting rates of climb for the appropriate airspeed as shown in Fig. 6-9 and determine the maximum point by drawing a smooth curve through your test points. Note that you may not have measured a point right at the peak, but a smooth curve should indicate about where this peak should occur. Also, note the velocity that falls directly under the peak. This is your best rate-of-climb speed, V_y.

Now, repeat this entire procedure for at least two other altitudes. Testing at a greater number of altitudes will result in more accurate final data. Then take the maximum values of rate of climb for each altitude and plot them against the respective altitudes as shown in Fig. 6-10. The altitudes, however, should be corrected to *density* altitude for this plot by one of the methods outlined in Chap. 1. The points should fall approximately on a straight line. Draw this line through the points and extend it until it touches both axes (zero lines) of your chart as shown.

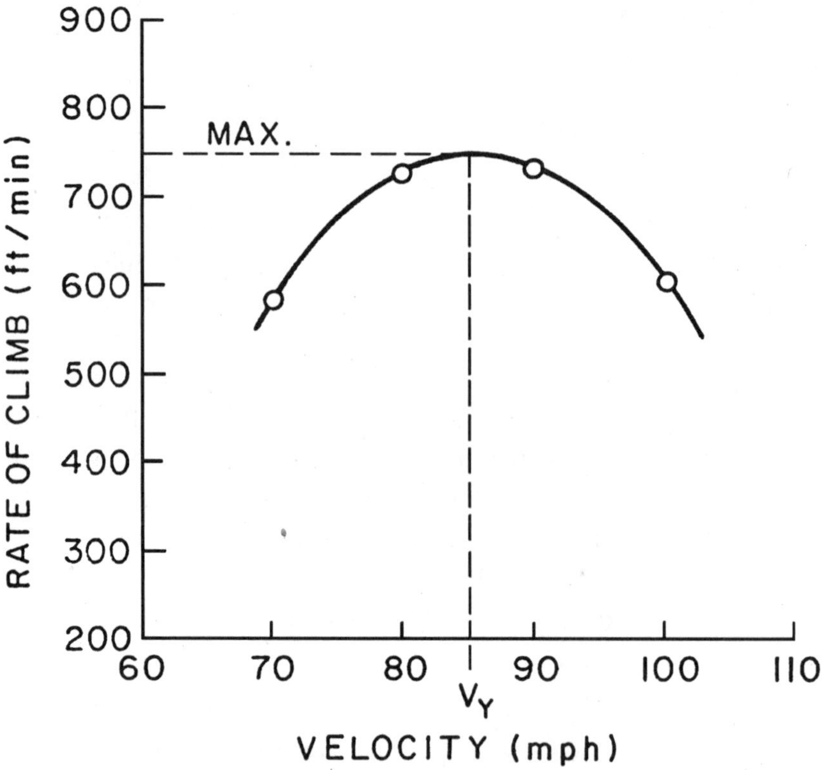

Fig. 6-9. Rate of climb versus velocity showing maximum point and best rate of climb speed.

The altitude where the rate of climb goes to zero is your absolute ceiling. The altitude corresponding to the rate of climb value of 100 ft/min is the service ceiling. The rate of climb for any density altitude can be obtained directly from this chart. Remember, though, to always calculate *density* altitude for your particular flight condition using the measured pressure altitude and temperature.

The best rate-of-climb speed for each altitude can also be plotted versus altitude. This speed should turn out to be nearly constant for all altitudes if measured in IAS or CAS. In terms of true airspeed, it will increase a bit with altitude. Such curves are shown in Fig. 6-11.

If the test airplane has retractable landing gear, you should measure rate of climb in both gear-up and gear-down configurations. The gear-down measurements need only be done at low altitudes to give an indication of initial climb ability right after takeoff. Twin-engine

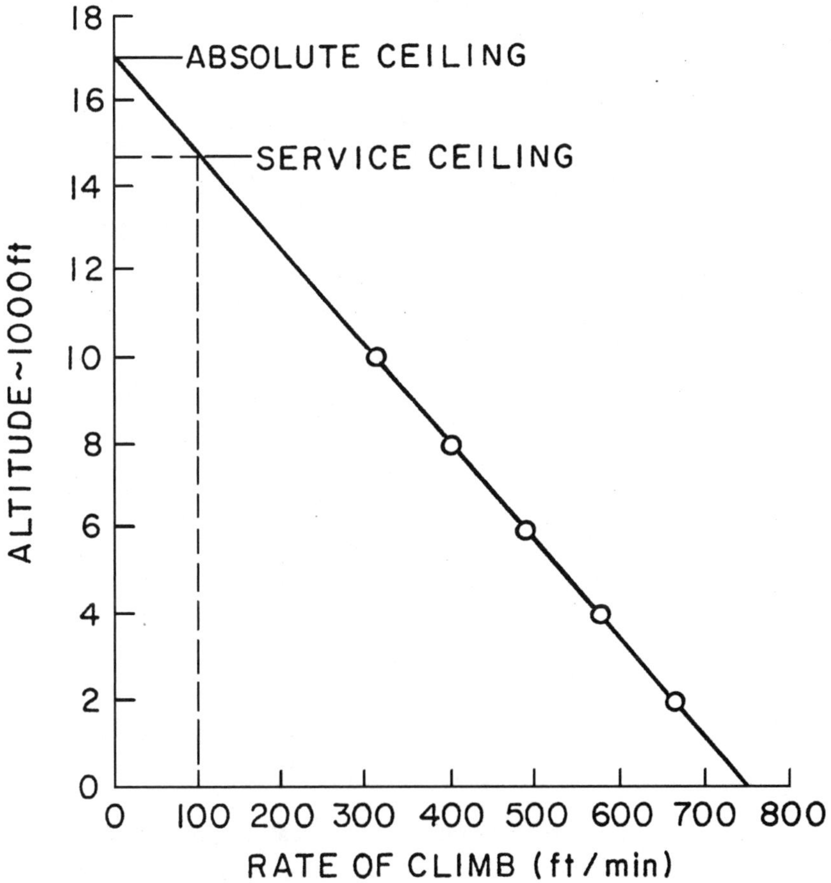

Fig. 6-10. Rate of climb versus density altitude showing determination of service and absolute ceiling.

airplanes should also be tested for single-engine climb performance. Make sure that a competent multiengine pilot is available for such tests and that all of the operating limitations for single-engine operation are observed.

Rate-of-Climb Test Example

- Test Conditions:
- Test altitude = 5000 ft (pressure)
- Initial altitude = 4800 ft (pressure)

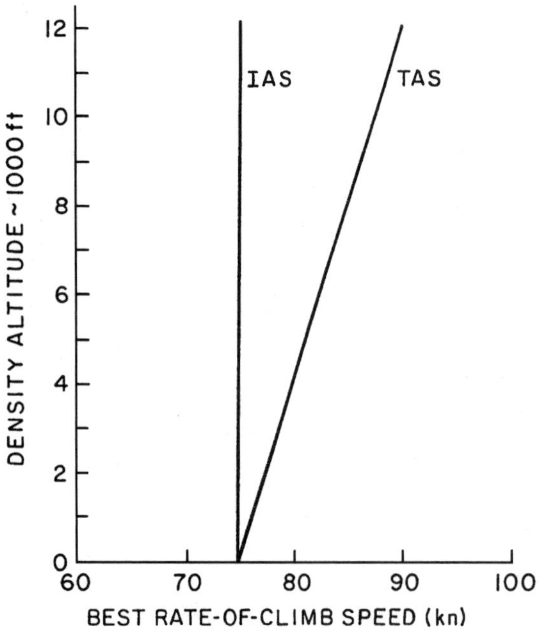

Fig. 6-11. Best rate-of-climb speed versus density altitude.

- Final altitude = 5200 ft (pressure)
- Time = 43 sec
- Temperature = 40°F
- Airspeed = 80 mph IAS

1. Determine rate of climb at this test point.

Rate of climb =

$$\left[\frac{\text{final altitude} - \text{initial altitude}}{\text{time}}\right](60) =$$

$$\left[\frac{5200 - 4800}{43}\right](60) = 558 \text{ ft/min}$$

2. Plot this point on rate of climb versus velocity graph (Fig. 6-12)
3. Assume other test points shown were obtained at this same altitude. Draw curve as shown. Maximum rate of climb is 600 ft/min.
4. From Fig. 1-3 correct 5000-ft pressure altitude to density altitude:

Density altitude = 5500 ft

5. Plot this point on altitude versus rate of climb chart (Fig. 6-13)

Time and Distance to Climb

Another consideration in climb performance is the time and distance to climb to cruise altitude. It wouldn't make much sense (or economy) to climb to an altitude that requires 20 min of climbing time, for example, if the overall flight only took 30 min. For short climbs, the time and distance could be calculated using an average value of rate of climb. However, when a climb to a fairly high altitude is planned, the rate of climb will change significantly from takeoff altitude to cruising level. In this case, using an average value of rate of climb would result in a significant error. For this reason it is best to actually measure the time to climb to various altitudes.The test for time to climb is fairly simple. You take off and immediately establish best rate of climb speed and maximum recommended climb power. For most airplanes, maximum climb power is the full power available. Some, however, may have certain revolutions per minute or manifold pressure limits for continuous operation. Check the flight handbook or engine operating instructions for these limitations. Also, lean the engine according to the engine manufacturer's instructions. Many engines require full-rich settings in climb up to a certain altitude and then leaning to best-power setting can be done above that altitude.

Set the altimeter to 29.92 and time the climb to various pressure altitudes. Start the stopwatch at takeoff and leave it running. Read the altitude for every minute (or every 30 sec) of elapsed time. Also, record the temperature at each reading and the total elapsed time. Continue this procedure up to, or close to, the maximum altitude at

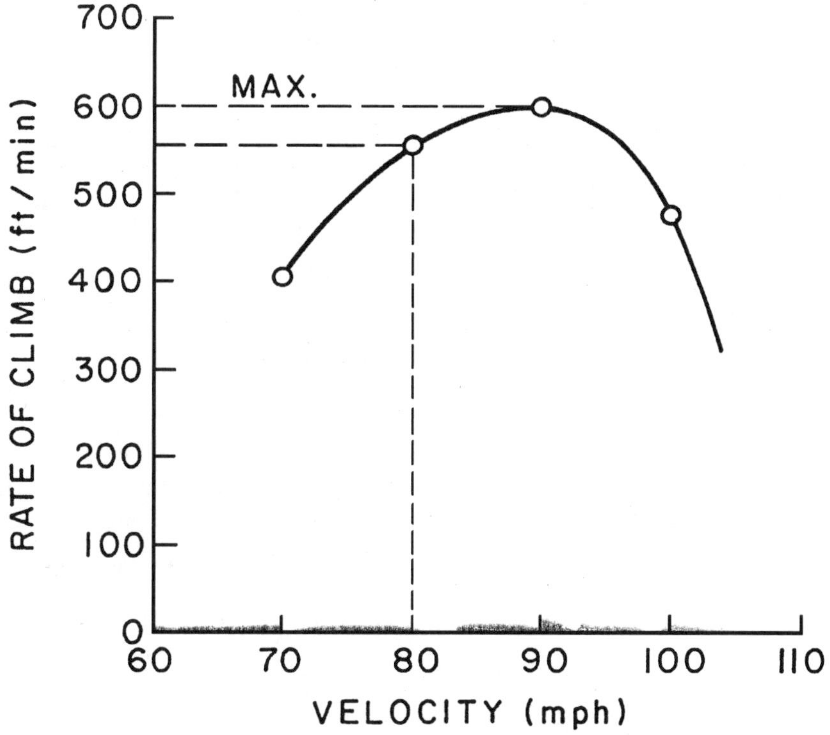

Fig. 6-12. Plotting rate of climb versus velocity and determining maximum.

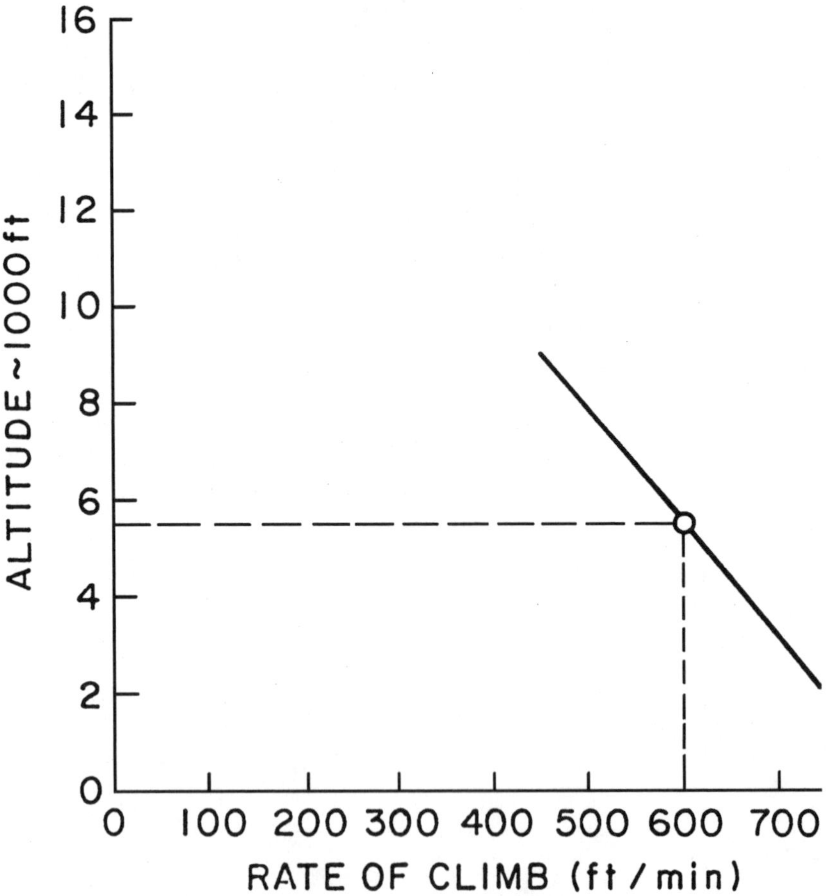

Fig. 6-13. Plotting maximum rate of climb for one altitude.

which you plan to fly. You may want to continue up to the ceiling. However, be sure to observe the special airspace restrictions that apply at high altitudes. VFR weather minimums increase above 10,000 ft and transponders are required above 12,500-ft MSL. Oxygen is also required above that level and recommended above 10,000 ft. For airplanes without such special equipment, it would be wise to limit climb tests to between 10,000 and 12,000 ft.

Using the pressure altitudes and corresponding temperatures recorded, convert the altitudes to density altitude values. Then simply plot the density altitudes reached against the respective times as shown in Fig. 6-14. One correction is necessary. Since your airport is probably not right at sea level, your climb will not start at zero altitude at time zero. This can easily be compensated for by extending the curve down to sea level. Since this point where the altitude is zero should occur at zero time, the whole curve can be shifted over to make it zero. The same amount of horizontal shift should be made to each point on the curve to yield a corrected curve, as shown. For example, if you took off at 1000-ft density altitude and the extended curve would cross the zero altitude line at −1 min, this would indicate that one minute would have been required to climb from sea level up to the 1000-ft level. Therefore, you should shift all of the curve to the right one minute. In this way the zero altitude and zero time points will coincide, the 1000-ft altitude will show +1 min of climb time required, and so on.

To determine the horizontal distance covered during climb, the time-to-climb data can be used in a calculation method. The best rate-of-climb speed (V_y) in TAS at various altitudes will also be required for this calculation. This information can be obtained from a graph similar to Fig. 6-11 obtained from your rate-of-climb tests. Take the average V_y

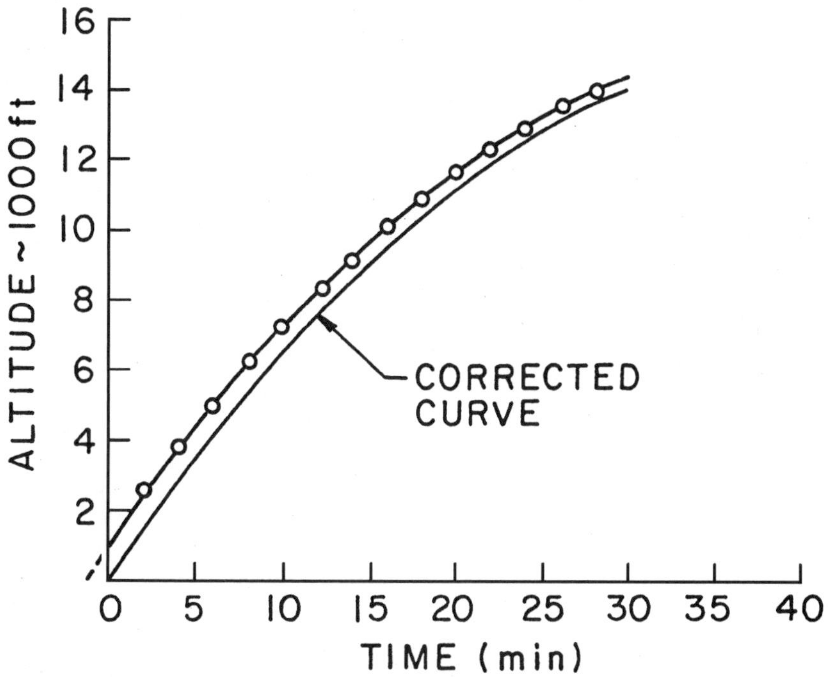

Fig. 6-14. Time to climb versus altitude showing correction for takeoff at standard sea level.

value between various altitudes and multiply it times the time required to climb between those altitudes.

$$d = V_{av}\frac{\Delta t}{60} \qquad \text{(for } \Delta t \text{ in minutes)}$$

As an example, assume the following data were obtained:

- 0–2000 ft:

$$\Delta t = 3.0 \text{ min}$$
$$V_y \text{ @ } 1000 \text{ ft} = 90 \text{ kn}$$

- 2000–4000 ft:

$$\Delta t = 3.3 \text{ min}$$
$$V_y \text{ @ } 3000 \text{ ft} = 92 \text{ kn}$$

$$d(0\text{–}2000) = 90\,\frac{3.0}{60} = 4.5 \text{ nm}$$

$$d(2000\text{–}4000) = 92\,\frac{3.3}{60} = 5.0 \text{ nm}$$

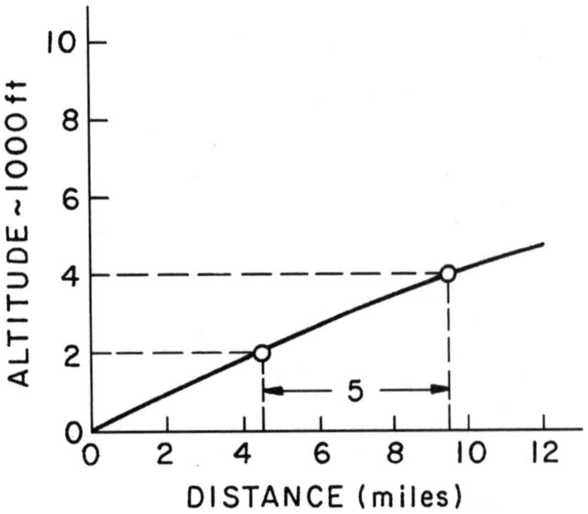

Fig. 6-15. Plotting distance to climb versus altitude.

In this case, the 4.5-mi distance would be plotted at the 2000-ft altitude point on the graph. The 5 mi would be added to this to yield 9.5 mi at the 4000-ft point and so on as shown in Fig. 6-15.

CHAPTER 7

Cruise Performance

In the cruise situation, the pilot is usually interested in knowing how long it will take to reach the destination. This time depends on the groundspeed, which, in turn, depends on the cruising airspeed. In addition, the pilot needs to know whether he has sufficient fuel to reach the destination, or if not, how far the airplane can go before it needs to be refueled. Cruise performance determination thus means determining the cruising speed and the range associated with that speed. In certain operations, such as traffic patrol or other localized flight activities, the endurance of an airplane is significant. Endurance is simply the time that an airplane can remain aloft rather than the distance that it can go in that time. For most operations, however, it is *range* that is of more importance to the pilot.

Power Determination

Both the cruising speed and the range depend on the power output of the engine. More power creates more thrust, hence, higher speed. More power also means higher fuel consumption, however, and therefore usually results in reduced range. Before addressing the specific problems of speed and range determination, we should have an understanding of power and how to properly select it.

Reciprocating engines have a certain horsepower rating, which is usually the maximum power that the engine can produce at sea level when operated within design limitations. Engines are not usually run at maximum power continuously, however. Most light airplane engines are designed to run normally at 75 percent of the rated brake horsepower (bhp). Power is usually specified in this way, that is, as a percentage of the rated brake horsepower. Thus, an engine rated at 200 hp would be delivering 150 hp at 75 percent bhp. The percentage applies to the rated horsepower and not the maximum power available at a certain altitude.

Of course, the engine can be run at less than 75 percent power and, consequently, consume less fuel. For most light airplanes, operation at less than 55 percent power results in very low cruise speeds. Thus, 55 percent is often regarded as the lower end of cruise power and is sometimes regarded as the *economy cruise* power setting. The upper limit, as mentioned, is usually 75 percent and referred to as *performance cruise* or *normal cruise* power. A good compromise, then, between the high fuel consumption at 75 percent and the low cruising speed at 55 percent is the middle range of 65 percent. Most airplane cruise performance charts will show curves for these three values of power. You can, of course, operate at points in between, or even below 55 percent. The three values of 75, 65, and 55 percent merely represent degrees of performance that are typically, high speed, intermediate, and economy values, respectively.

Before proceeding to determine performance at certain power settings, we must know how to obtain a specific amount of power from the engine. For this purpose, it is necessary to consult the aircraft operating manual. If your test airplane is not a standard production model or has a modified engine installed, then it will be necessary to contact the engine manufacturer for power data. The manual for a production aircraft will provide power setting information such as is shown in Fig. 7-1. This chart is really for a Piper Cherokee 140 and shows, for example, that at sea level, 2375 rpm is the setting for 75 percent power. This airplane has a fixed pitch propeller which means that, at higher altitudes, a higher rpm is required to obtain the same amount of power, due to the lower air density. Therefore, at 4500-ft density altitude, a setting of 2500 rpm is necessary to obtain 75 percent power.

If the airplane has a constant speed propeller, the pilot has more of a choice in power settings. Such airplanes have both a propeller control (to set revolutions per minute) and a throttle (to set manifold pressure). Various combinations of these two controls can be used to obtain the same percentage of power. Table 7-1 is from the Piper Arrow handbook. It shows the combinations of revolutions per minute and MAP for various power settings and altitudes. For example, to obtain 65 percent power at 3000 ft, you could select either 25.1 in of MAP and 2100 rpm or 22.2 in and 2400 rpm. In either case, with the engine developing the same amount of power, the airspeed and the fuel consumption should be the same. It might be helpful, for flight test purposes, to plot such data in chart form if your handbook presents it in a table such as this one. In any case, you must have

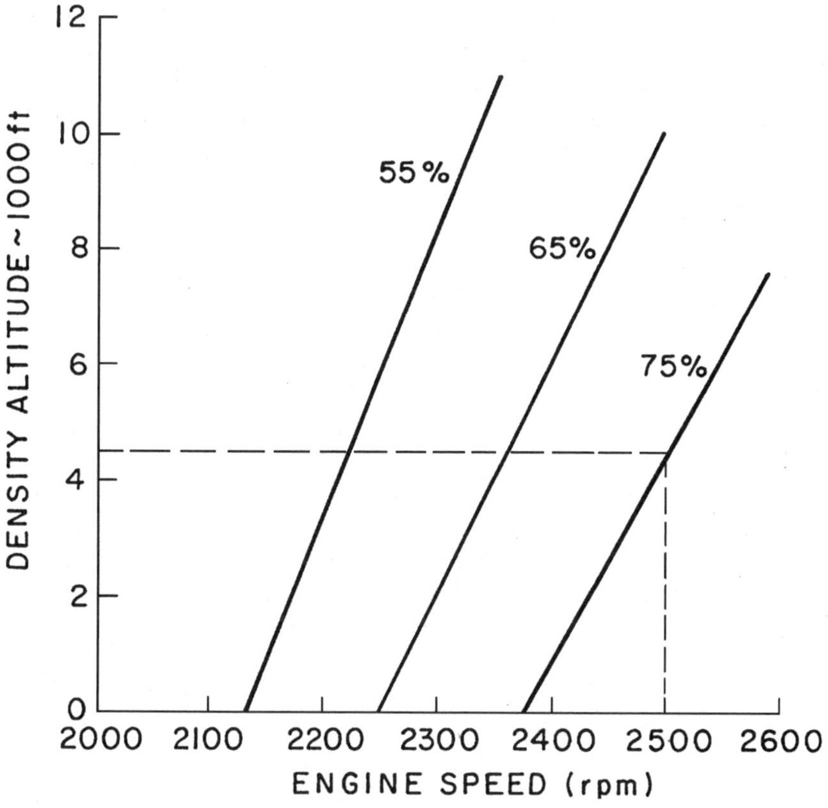

Fig. 7-1. Power chart for Piper Cherokee 140 showing RPM for various percent power at various altitudes. *(Courtesy of The New Piper Aircraft, Inc.)*

Table 7-1. Power Setting Data for the Piper Arrow II

POWER SETTING TABLE—LYCOMING MODEL 10-360-C SERIES, 200 HP ENGINE							
		110 HP—55% RATED RPM AND MAN. PRESS.		130 HP—65% RATED RPM AND MAN. PRESS.		150 HP—75% RATED RPM AND MAN. PRESS.	
PRESS. ALT, FT	STD. ALT TEMP, °F	2100	2400	2100	2400	2400	PRESS. ALT, FT
SL	59	22.9	20.4	25.9	22.9	25.5	SL
1,000	55	22.7	20.2	25.6	22.7	25.2	1,000
2,000	52	22.4	20.0	25.4	22.5	25.0	2,000
3,000	48	22.2	19.8	25.1	22.2	24.7	3,000
4,000	45	21.9	19.5	24.8	22.0	24.4	4,000
5,000	41	21.7	19.3	FT	21.7	FT	5,000
6,000	38	21.4	19.1	—	21.5	—	6,000
7,000	34	21.2	18.9	—	21.3	—	7,000
8,000	31	21.0	18.7	—	21.0		8,000
9,000	27	FT	18.5	—	FT		9,000
10,000	23	—	18.3				10,000
11,000	19	—	18.1				11,000
12,000	16	—	17.8				12,000
13,000	12	—	17.6				13,000
14,000	9	—	FT				14,000

To maintain constant power, correct manifold pressure approximately 0.16 in Hg for each 10°F variation in inlet air temperature from standard altitude temperature. Add manifold pressure for air temperatures above standard; subtract for temperatures below standard.

Source: Courtesy of The New Piper Aircraft, Inc.

information which tells you what power setting you need to develop a certain brake horsepower at a certain density altitude.

One other control which has a significant effect on the engine power output is the fuel/air mixture control. Performance that depends on engine power is usually indicated as being for a specific mixture setting. It is important for cruise performance that a certain mixture be maintained throughout the tests. This is best done by use of an exhaust gas temperature gauge (EGT). There are two mixture settings normally used in determining performance. One is known as "best economy mixture" and usually corresponds to the peak point on the EGT gauge. This mixture gives the lowest fuel consumption, but at a little sacrifice in power. The other setting is "best power mixture" which is the setting that gives the maximum engine power. It is usually achieved by backing off the mixture control to about 100°F below the peak temperature. This setting can also be achieved without an EGT by leaning until a power loss is noted (sometimes accompanied by roughness) and then enriching until the engine just regains power and runs smoothly. This procedure, known as "normal leaning" is not recommended for fuel-injected engines. For any engine, use of the EGT gauge is best for flight testing.

The engine manufacturer's instructions should be followed closely in leaning procedures. Most engines should not be leaned at all above 75 percent power or above 65 percent in the case of geared engines. Since many engines will not develop 75 percent power above 5000 ft, flight manuals sometimes indicate not to lean below that altitude. You can really lean at any altitude as long as you know what you are doing. By that we mean knowing what your power setting actually is. If it is 75 percent or below for conventional engines, or 65 percent for geared engines, leaning can be accomplished safely.

Cruising Speed Test Procedure

Very basically, the cruise speed test simply involves flying at a certain power setting and reading the resulting airspeed. There are certain techniques that must be applied, however, in order to get accurate and meaningful data. Tests must be run at different altitudes to get the variation in airspeed with altitude. It is also desirable to run tests at various power settings, such as 55, 65, and 75 percent.

Begin the tests at as low an altitude as safe flight can be conducted. It would be desirable to start right at sea level if it were possible. Since this is not a very realistic situation, the lowest practical altitude can be used. It does not matter what exact altitude you select. You must, though, determine what the *density* altitude is at that particular true altitude. This is done by measuring temperature and pressure altitude and then using a calculator or the chart in Fig. 1-4 to get density altitude. Now you can go to your operator's manual to see what revolutions per minute (or revolutions per minute and MAP) you need to obtain certain power values at that density altitude.

Set up the power for 55 percent and fly straight and level. It is particularly important for this test to be done in very smooth air. Another trick in getting good speed readings is to fly for at least a minute, or better yet, several minutes at one setting. Make sure that no climb or descent tendency exists. When you are certain that the airplane is stabilized in level flight and the airspeed reading is holding constantly, take the airspeed reading. Then repeat the procedure for 65 percent and 75 percent, or other settings if desired. It would be advisable to run tests at least at 55 percent and 75 percent. The number of other settings chosen is up to you.

After completing tests for all settings at the first altitude, climb about 2000 ft and repeat the entire procedure. Do this for at least three different altitudes (or preferably, a few more). Eventually you will reach an altitude where you cannot obtain 75 percent power, but continue the tests at the lower power settings. If you go high enough, you will find that you will not be able to achieve 65 percent above a certain height. At this point you have probably gone high enough to obtain sufficient data for the entire test. Don't forget to always calculate the *density* altitude for each level tested.

Since cruising speed information will eventually be used to determine how fast the airplane is moving toward its destination, the speed values will have to be in *true* airspeed. Thus, it is necessary to convert all of the indicated airspeed readings to true. For this reason it is necessary to record the temperature and pressure altitude for each test point. Convert the IAS reading first to CAS by the calibration curve obtained from the test in Chap. 3. Then, correct CAS to TAS for the appropriate temperature and pressure altitude, either with a flight computer or by use of the density ratio and the equation in Chap. 3.

The resulting values of true airspeed can now be plotted against density altitude for the various power setting values tested, as shown in Fig. 7-2. The variation with altitude for a given power setting should follow a straight line. Some error in the testing is bound to occur, however, and the points will probably not fall exactly in line. Fit a straight line to the points as closely as possible. Also extend the line on down to sea level to determine what the cruise speed would be there. Do this for each power setting tested.

Maximum Speed Test Procedure

Although cruising is of most importance in cross-country flying, it is also desirable to know the maximum speed that could be attained at

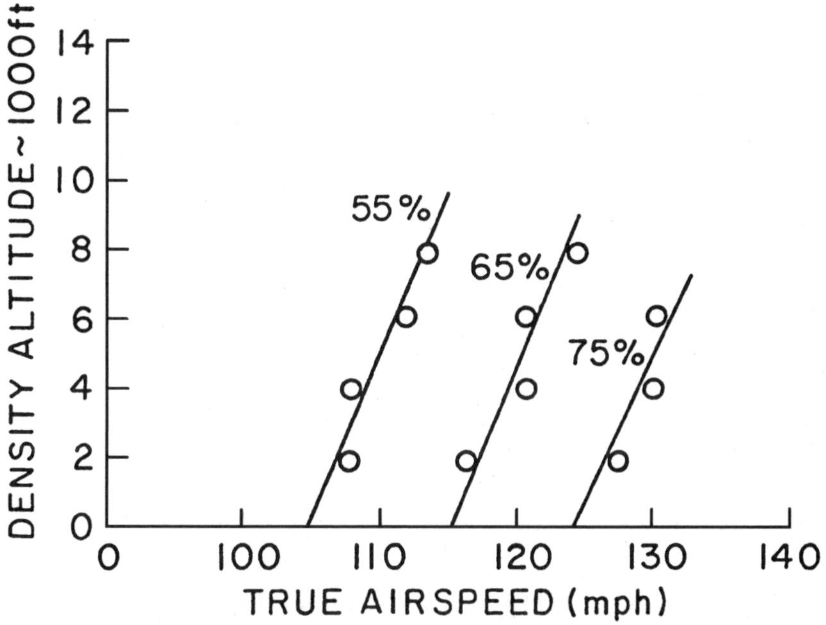

Fig. 7-2. Typical plot of cruise speed versus altitude at various power settings.

any altitude. Since maximum power is attainable at sea level, due to the highest air density here, the maximum possible level flight speed also occurs at sea level. The maximum speed then decreases with altitude as the density decreases. A plot of maximum speed against altitude is shown in Fig. 7-3.

Determining maximum speed is done by the same procedure as utilized for cruising speed, except that full power is applied rather than some percentage of rated power. As before, it should be measured at various altitudes and the respective values of density altitude determined (Fig. 7-3). For maximum speed it is especially important to establish a true level flight condition and to hold that attitude for several minutes before taking the airspeed reading. Only the very slightest variation in angle of attack will cause a vertical movement and an erroneous maximum speed indication.

Caution should be used in operating at full power at lower altitudes. Most engines can be run continuously at their rated power without damage; however, engine wear may greatly increase above 75 percent. Other engines have a "takeoff power" rating which means

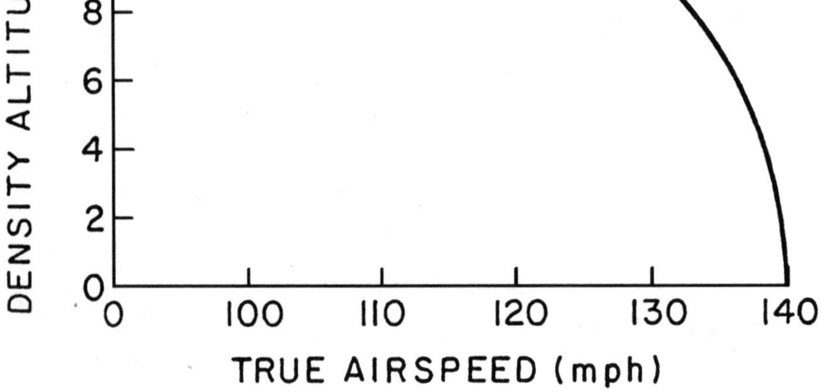

Fig. 7-3. Maximum airspeed versus altitude.

that they can be operated at such power only for short durations of time, as would be done in takeoff. Most of the lower horsepower four and six cylinder engines can be run at maximum power indefinitely as long as engine temperature is within limits. If you are uncertain of the engine limitations, consult the manufacturer. If this is not possible, it is best to limit the maximum speed tests to higher altitudes where more than 75 percent power is not attainable. The approximate maximum altitude for 75 percent power should be apparent from the cruise power tests.

Determine the maximum speed at a number of density altitudes and convert the airspeed values to TAS. Then, plot these points on the cruise speed chart, as shown in Fig. 7-4. A curve fitted to these points should then form the upper limit of your cruise speed lines.

Since 75 percent power is usually specified as maximum cruise power, the intersection of the 75 percent line with the maximum speed curve marks the overall maximum cruise speed point, the altitude for maximum cruising true airspeed. Above this altitude, the maximum cruise and the maximum (full power) speed are the same. Below this

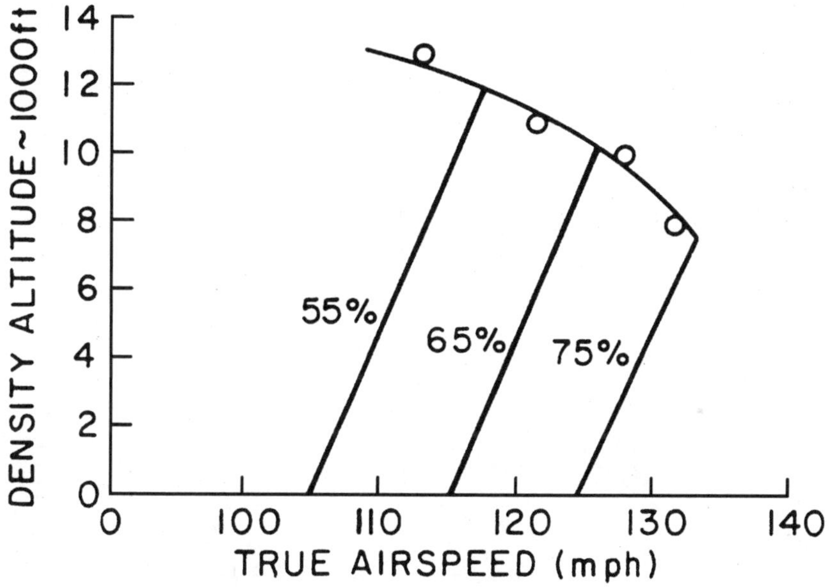

Fig. 7-4. Complete speed envelope constructed from cruise speed and maximum speed tests.

altitude, the maximum cruise speed is that for 75 percent power. This situation is apparent from Fig. 7-5. The extension of the maximum speed line below the max cruise altitude (shown as a dashed line) represents an area of *possible* speeds, but restricted due to engine design limitations.

Note that some airplanes with constant-speed props have a maximum recommended cruise revolutions per minute. The maximum practical speed for such aircraft would be determined by setting the prop control for this revolutions per minute value and then establishing full power by application of full throttle. Such procedure performed at various altitudes would then establish an upper limit on *cruising* airspeeds for a specified value of revolutions per minute.

Range

The range, or distance that an airplane can fly, depends on its cruising speed, its fuel capacity, and its rate of fuel consumption. Specifically,

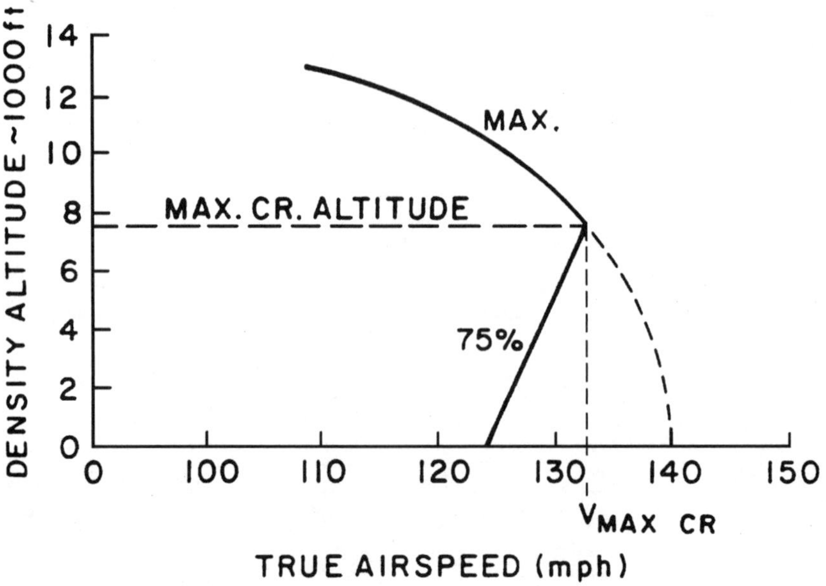

Fig. 7-5. Determination of optimum cruising altitude and maximum cruising speed.

it is determined as the speed times the fuel capacity divided by the fuel consumption. In equation form this is given as,

$$\text{Range} = \frac{V_{\text{cruise}}(\text{fuel})}{(\text{fuel consumption})}$$

For range in miles, the speed would have to be in miles per hour, the fuel quantity in gallons (or pounds), and the consumption rate in gallons (or pounds) per hour.

Range, however, varies considerably for different power settings. We have already seen in the section on cruising speed how speed varies with different power settings at different altitudes. The fuel consumption also varies with power setting. In fact it is pretty much directly proportional to the brake horsepower output of the engine. Since both speed and fuel consumption affect the range, it should be obvious that power has a great deal to do with determining range. At high power we get higher airspeed and this tends to increase range. However, high power requires high fuel consumption, and this term, being in the denominator of the range equation, will reduce range as it gets larger.

Figure 7-6 shows the power-required curve for a Piper Arrow II. This curve was constructed from flight test data. Notice that in the normal cruising range (55 to 75 percent power) that slower cruising speed requires a smaller amount of power. Less power is required because the drag drops off at lower speeds. Also, note that the power required drops off at a greater rate than the velocity in this area. Therefore, the overall result is an increase in range. This trend continues down to the point of minimum drag. Below this point, the velocity drops off at a greater rate than power, and range is *reduced*. Therefore, the point on the power curve where the drag is a minimum represents the power setting for *maximum* possible range.

Notice that the speed at this point of maximum range power is only 112 mph. This is pretty slow for an airplane capable of more than 150-mph cruise. We invest in airplanes with retractable gear and constant-speed props to attain some degree of cruise performance. At this rate you may as well have taken a trainer. The fact is that very seldom does anyone ever fly at maximum range speed. It is more of an academic

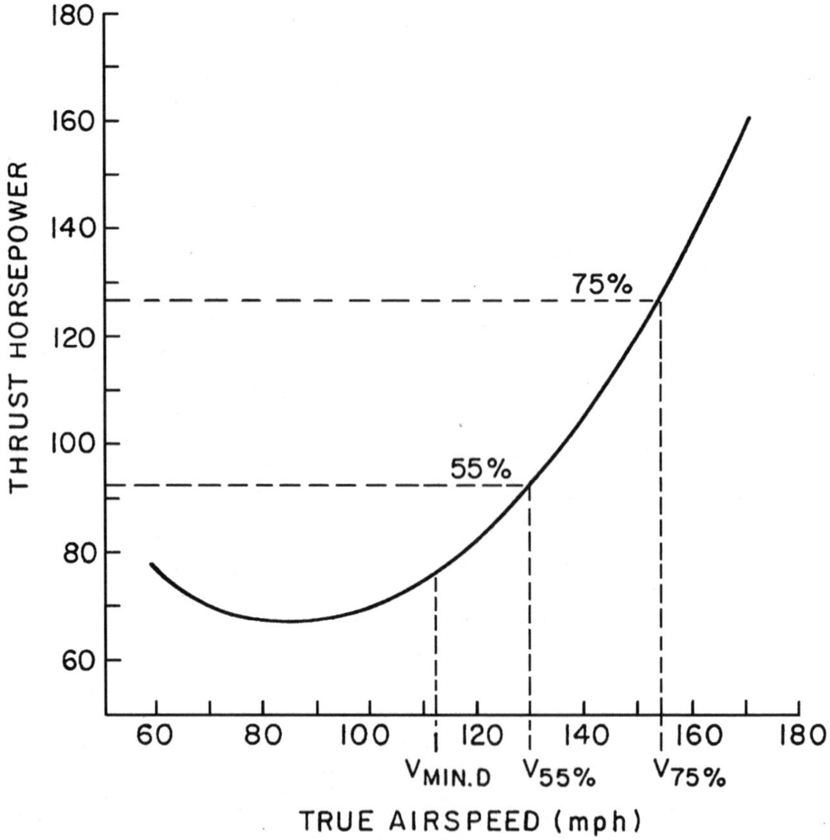

Fig. 7-6. Power required curve constructed from flight tests on Piper Arrow II showing power required at various level flight speeds.

consideration. We have already discussed, in the section on cruising speed, that most flying is done at power settings of between 55 percent and 75 percent brake horsepower. In this power range, the Arrow cruises from about 130 to 153 mph. However, note that there is a considerable decrease in power required at 55 percent over that at 75 percent power. Thus, there would be a significant increase in range at the lower power setting.

Another quality of the airplane that affects range is the gross weight. The power curve, as shown, is really for one particular gross weight. At a lower gross weight the induced drag is less; therefore, a slightly higher speed is attainable at the same power setting. This fact complicates the exact determination of range, since weight is continually changing in flight. As fuel is consumed the airplane gets lighter. If fuel accounts for a major portion of the gross weight, the velocity corresponding to a certain power value could vary significantly. To calculate range, then, from the equation given above, you would have to use various values of velocity for various increments of weight to get increments of range, and then sum them for total range. For many airplanes, though, the fuel weight is a small portion of the gross weight and, in this case, an average value of velocity at average weight is sufficient. This is particularly true of many light single-engine airplanes whose total fuel weight is about 12 to 15 percent of the gross. For such airplanes, the velocity values obtained for various altitudes and power settings in the cruise speed tests are adequate to be used as constant values in calculating range.

Range Test Procedures

Just as the takeoff test does not require making an actual takeoff, range determination does not require flying the entire range of the airplane. Flying until the fuel is completely exhausted would be a bit foolhardy, anyway. Range is best determined by calculation from the equation given above. Testing is involved to determine the various terms in the equation. The amount of usable fuel should be known from the airplane operator's handbook. If the airplane is experimental, the actual capacity of the tanks can be measured when filling them. The cruise velocity for various altitudes and power settings should already be known from the cruise speed test.

The remaining item to determine is the fuel consumption. This determination will be the objective of the flight tests involving range. Fuel consumption rate is essentially constant for a given power setting. It could be measured at almost any altitude as long as the desired power is properly set. The power setting should be determined in flight just as it was for the cruise speed tests. Measure the pressure altitude and the temperature at your flight level and calculate density altitude either with a computer or the chart in Fig. 1-4. Then, from the airplane manual, obtain the power settings in revolutions per minute (or revolutions per minute and MAP for constant-speed prop) at that density altitude. Fly the airplane at the proper settings for 55 and 75 percent and any other settings for which data are desired.

An important step in setting up power to determine fuel consumption is to properly adjust the mixture control. Any desired mixture can be chosen for the tests, but then, it should be used consistently throughout. Most production airplane manuals use best economy mixture setting in determining range and best power mixture for cruising speed. Since cruise speed is measured with best power mixture, range could also be determined at this same setting. Best power mixture is also attainable without an EGT gauge (Fig. 7-7). Remember, though, that somewhat lower fuel consumption and slightly longer range could be achieved at best economy setting.

Fig. 7-7. Typical exhaust gas temperature (EGT) gauge, which is very useful for determining mixture setting.

At each power setting, measure the fuel flow. This can be done in several ways. The best way is to temporarily install an accurately calibrated fuel flowmeter in the fuel line. It will read directly and instantaneously in gallons per hour (or pounds per hour). Most airplanes with fuel-injected engines will have a fuel flowmeter installed in the panel. Such a meter can be used for this purpose, but many of these do not score too highly in the accuracy column (Fig. 7-8).

Without an accurate fuel flowmeter, another method has to be resorted to. This is to fly the airplane at a constant power setting for a significant amount of time and then measure the fuel necessary to refill the tank (Fig. 7-9). This procedure is fairly straightforward if your airplane has more than one fuel tank and each can be selected individually. In this case you make your takeoff, climb, descent, and landing on one tank and run on the other only when you are cruising with an accurate power setting. Keep the tank that is to be measured completely full until you have a definite amount of cruise power established. Then switch to that tank and fly for about an hour or two

Fig. 7-8. Fuel consumption determination can be made directly from the fuel flow gauge.

Fig. 7-9. By measuring fuel consumed from fuel pump meter during refueling, consumption can be calculated.

or whatever the tank capacity will permit. Before descent and landing, switch over to the original tank.

After landing, carefully monitor the refueling to determine the exact amount of fuel needed to replenish the cruise tank. Most flow indicators on fuel pumps and trucks are reasonably accurate. If you doubt the accuracy, you might have a gallon or two pumped into a measuring container and check it against the reading on the pump. A more exact way would be to pump all of the fuel into an accurate liquid measure and then pour it into the tank. This is a bit time-consuming and may try the patience of the line attendant. Use a bit of discretion in selecting an airport for your refueling. The FBO at LaGuardia or Washington National may not be too cooperative in this kind of procedure while he has three Learjets and a Sabreliner waiting to be refueled.

The fuel consumption can then be determined by dividing the numbers of gallons by the hours flown:

$$FC = \frac{\text{gallons}}{\text{time (hr)}}$$

Note that the consumption rate at only one power setting can be measured on each flight if done in this manner. Another flight is required for each additional power value desired. Since several hours of flying time are involved for each of these tests you may as well combine it with a planned trip. It might even be a good excuse to make a trip. One power setting can be used on the way to the destination and another on the return trip. A three-legged course could also be planned for three power settings. It does not matter what cruise altitude you select, as long as you determine the settings necessary for a specific percentage of power at the altitude flown. The fuel consumption is essentially constant for a given power setting, regardless of other conditions.

If your airplane has only one fuel tank or, if individual tanks are not selectable, the procedure is a little more complicated. However, all is not lost. The test can still be done, although somewhat less accurately. The trick here is to use a given percentage of power for as much of the flight as possible. Switch to your desired power setting as soon as possible after takeoff and sufficient altitude have been achieved. Then retain that setting until just before landing. Remember that you can climb at 75 percent and other reduced power settings. You can also descend by increasing airspeed above that necessary for level flight. A long, shallow descent is probably the most difficult part of the maneuvers for this method. This can be done, however, right down to the traffic pattern. Again, choose your destination airport carefully. Don't plan such a test during an IFR flight into Class B Airspace.

The important point in the single tank method is to keep the operations at other than cruise power settings to a small percentage of the overall flight. This procedure necessitates a fairly long flight. You must also have a pretty good idea of the power setting for your chosen power value right after takeoff. Then as you climb to a cruising altitude, you will have to continually reset your throttle for the same power at various altitudes. This is a somewhat demanding task and usually requires a crew of at least two. A preliminary flight could even be made to measure temperature and pressure at various altitudes. The exact setting for each altitude could then be determined and after filling the tanks, the actual test run could be made utilizing these settings. This variation of the basic method could be handled adequately by one person. Remember, also, to lean to the desired mixture setting at any altitude involved in the test.

Regardless of what exact method was used, a fuel consumption figure for each power setting should now be established. Obtain cruise velocities for various power values and altitude from the chart constructed from the cruise speed tests. This chart should be similar to that shown in Fig. 7-4. Apply various values of speed and the proper fuel consumption value to the equation for range:

$$R = \frac{V_c(\text{fuel})}{FC}$$

It is best to read velocities from the lines on the speed chart rather than the exact measured points. The straight line curve in the speed chart tends to average out velocities and minimizes the error involved in measuring individual points on the curve.

Calculate the range for several altitudes for a given power setting, say 55 percent, and then repeat for another setting, such as 75 percent. Plot these points on a range versus altitude chart and draw a straight line through each set of points as shown in Fig. 7-10.

This chart now represents what is known as *absolute* range, if you have used the total fuel capacity in the equation. This range does not allow for climb and descent nor does it account for any amount of fuel reserve. Obviously, then, it does not represent an achievable range, but only a maximum limit. The actual distance you could safely fly, under no-wind conditions, would be somewhat less. Range charts often also include curves for range with a 45-min reserve. The 45-min reserve is determined as the distance that could be flown in 45 min at 55 percent power. This distance is easily determined by multiplying the speed at the proper altitude and 55 percent power by 0.75 (since 45 minutes is 0.75 hours):

$$d_{\text{reserve}} = 0.75\ V_{55\%}$$

This distance is then subtracted from your absolute range at that altitude to yield the range with 45 min reserve. This same distance is subtracted from the 75 and 65 percent range curves as well, since the reserve distance is always considered to be flown at 55 percent power.

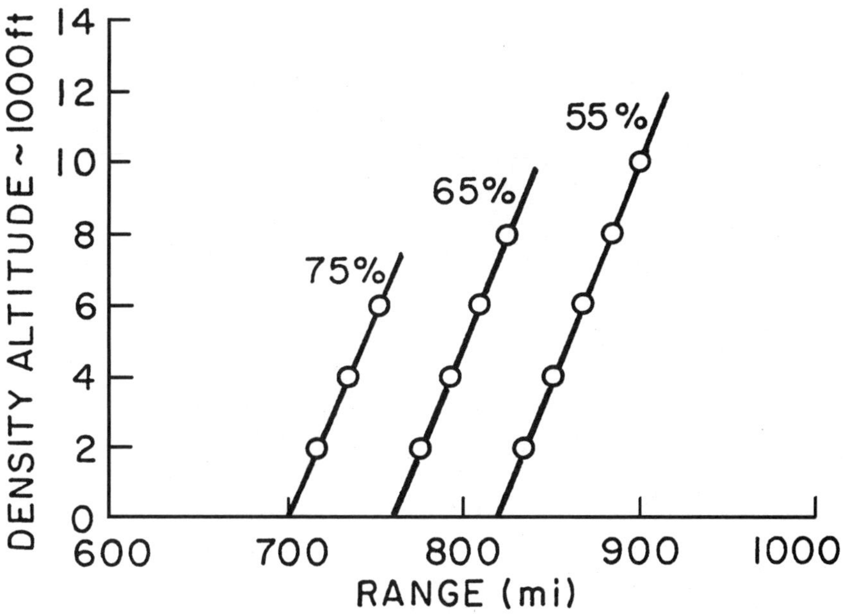

Fig. 7-10. Typical chart of range versus density altitude for various power settings as constructed from test data.

Endurance

For some operations it is desirable to know the amount of time that the airplane can remain airborne rather than the distance it can fly. This time is known as *endurance.* This figure is important, for example, if the pilot is in a holding pattern awaiting landing clearance. The endurance for a certain power setting is easily obtained by dividing the range at that setting and a certain altitude by the cruise speed at the same conditions:

$$E = \text{endurance} = \frac{R}{V_c}$$

Endurance is essentially constant for all altitudes. It also increases, like range, with decreased power. The maximum endurance is obtained at the very lowest point on the power curve. From Fig. 7-6, the speed at this power for the Piper Arrow is about 85 mph. Note that this speed is significantly lower than the maximum range speed.

Cruise Performance Example

Problem: Find cruise speed and range at 75 percent power.

Test condition:

Pressure altitude = 5000 ft

Temperature = 55°F

1. Determine density altitude. From Fig. 1-3, density altitude is 6000 ft

2. Refer to power chart for the airplane to get revolutions per minute for 75 percent power. Read 2620 rpm.

a. Set 2620 rpm, lean to best power mixture, and read velocity and fuel flow:

$$V_i = 110 \text{ kn}$$

$$FC = 10.25 \text{ gal/hr}$$

4. Correct IAS to TAS. Assume negligible error in the airspeed system so that IAS = CAS. From Fig. 1-4, ρ/ρ_0 at 6000 ft is 0.836:

$$\text{TAS} = \frac{1}{\sqrt{\rho/\rho_0}}(\text{CAS}) = \frac{1}{\sqrt{.836}}(110) = 120 \text{ kn}$$

5. Determine the range at this point (75 percent power and 6000 ft). Fuel capacity is 48 gal:

$$R = \frac{V(\text{fuel})}{FC} = \frac{120\ (48)}{10.25} = 562 \text{ nm}$$

CHAPTER 8

Descent and Landing Performance

In Chap. 6 we discussed how climb results when the power available is more than the power required. The rate of climb was shown to be proportional to the excess power available. If the power is reduced to below what is required at a particular airspeed, then a negative rate of climb value would result. Negative rate of climb is referred to as rate of sink, RS. The airplane would *descend* in such a condition proportional to the difference between power available and power required. Since the power available is less than that required, it could be referred to as a "decrement" in power rather than an excess, as in the climb case.

The maximum rate of sink, of course, occurs when the power is reduced to zero. The power available curve in Fig. 6-8 is shown at its maximum location and it was noted that the power controls could be adjusted to move this curve anywhere between maximum and zero power. If it were reduced to zero, then the power available would be the axis of the graph or line of zero thrust horsepower. The rate of sink in this condition is now proportional to the difference between this axis and the power required curve at any velocity, or to this decrement in power. Figure 8-1 shows this situation, which actually represents a power-off glide. Note that at the velocity for minimum power, the decrement of power would be smallest and, hence, the rate of sink

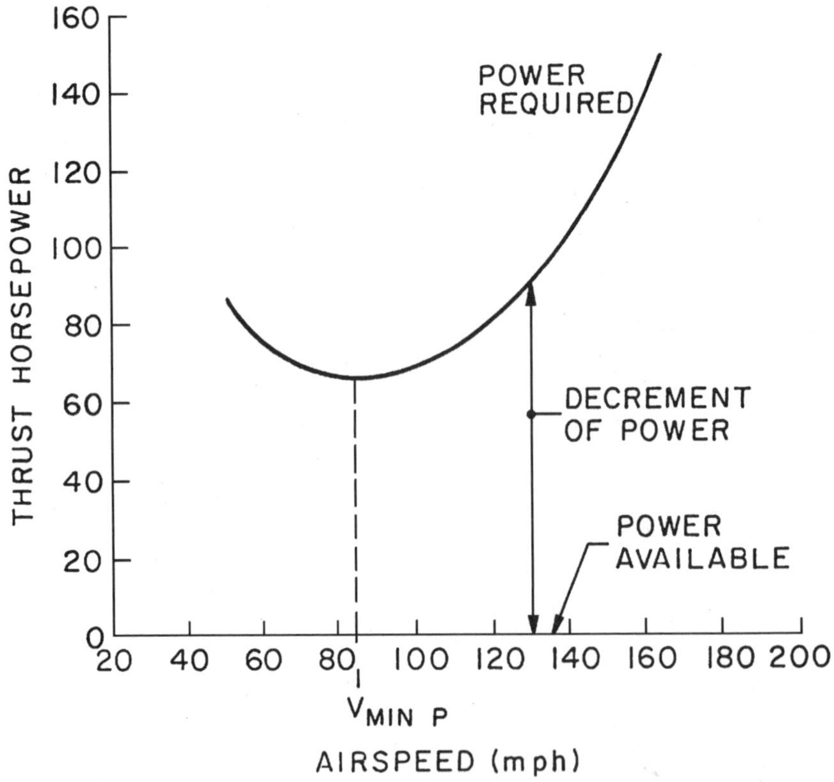

Fig. 8-1. Decrement of power resulting from zero power available as in power-off glide.

would be lowest. As the speed increases or decreases from this point, the sink rate would increase. Gliding at the speed for minimum power would, thus, give you the longest *time* in a glide.

Power-off glide performance is of most concern in the event of an actual engine failure, in which case you really do have zero power available. If this situation should occur, it is usually the maximum glide *distance* that the pilot is interested in rather than glide time. The gliding distance possible from any given altitude depends on the ratio of the horizontal (or forward) speed to the vertical speed (or rate of sink). The higher this speed ratio the greater will be the gliding distance.

The maximum horizontal-to-vertical speed ratio could be determined if you had a plot of the power required curve. A line could be drawn form the origin (the point where both thrust horsepower and *V* are zero) to where it just touches, or becomes a tangent, to the power curve, as shown in Fig. 8-2. At the point where it just touches is the maximum *ratio* of *V* to RS. This ratio is also referred to as the glide ratio, GR. It turns out that glide ratio is the same as the lift-to-drag ratio (*L/D*). Therefore, this point of maximum horizontal-to-vertical speed is the maximum *L/D* point of operation and gives the maximum glide distance. You, of course, achieve this situation by flying at the velocity corresponding to this point on the velocity scale.

The distance, then, that one can glide is the glide ratio (or *L/D* ratio) at any particular speed being flown times the altitude (or vertical distance) from which the glide is started:

$$d_g = (L/D)h$$

The maximum glide distance would be achieved when flying at maximum *L/D*. Remember that the altitude here is the actual height above the surface and not above sea level. This relationship only holds for

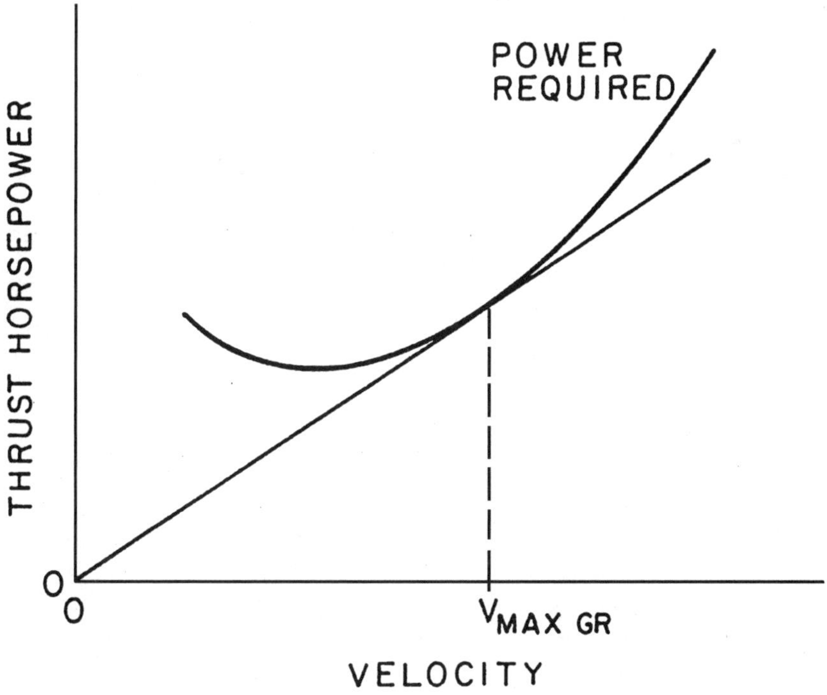

Fig. 8-2. Location of speed for maximum glide ratio on curve of power required versus velocity.

normal glides in which the glide angle is relatively small. In this case, the horizontal speed of the airplane is just about the same as the forward speed. This is not true in the case of a steep, screaming dive.

Determining Glide Ratio

The glide ratio is determined by measuring the rate-of-sink at various airspeeds. This is done much like the rate-of-climb determination except that a power-off glide is established rather than a full power climb. The time to descend through an increment of altitude is measured, and rate of sink determined by dividing the altitude change by time:

$$\mathrm{RS} = \frac{\Delta h}{t}$$

This procedure is repeated for various airspeeds and these speeds are recorded. The glide ratio is, then, the speed divided by the respective rate of sink at that speed:

$$\mathrm{GR} = \frac{V}{\mathrm{RS}}$$

Both horizontal and vertical speed must be in the same units. Rate of sink is probably measured most easily in feet per second. Therefore, forward speed will have to be converted to feet per second. If V is recorded in mph, the equation is:

$$\mathrm{GR} = \frac{1.467(V)}{\mathrm{RS}}$$

If V is recorded in knots, then the glide ratio equation is:

$$\mathrm{GR} = \frac{1.687(V)}{\mathrm{RS}}$$

The glide, or L/D, ratio does not change with altitude, so that it does not matter what altitude is selected for the experiments. The same velocity will also be associated with the same L/D as long as it is measured in *calibrated* airspeed (CAS). Velocity should be recorded in IAS and then corrected for airspeed error, if significant, to CAS. To get an accurate ratio of forward distance to vertical distance, the airspeed must be further corrected to true airspeed for use in the above equations. Correction from CAS to TAS requires the density ratio. Pressure altitude and temperature, therefore, must be recorded in order to determine the density ratio.

Plot the resulting *L/D* ratios versus the calibrated airspeed to yield a curve such as shown in Fig. 8-3. You may want to plot this in IAS, since that is the value actually read on the indicator. It does not usually differ greatly from CAS, and is somewhat proportional. The

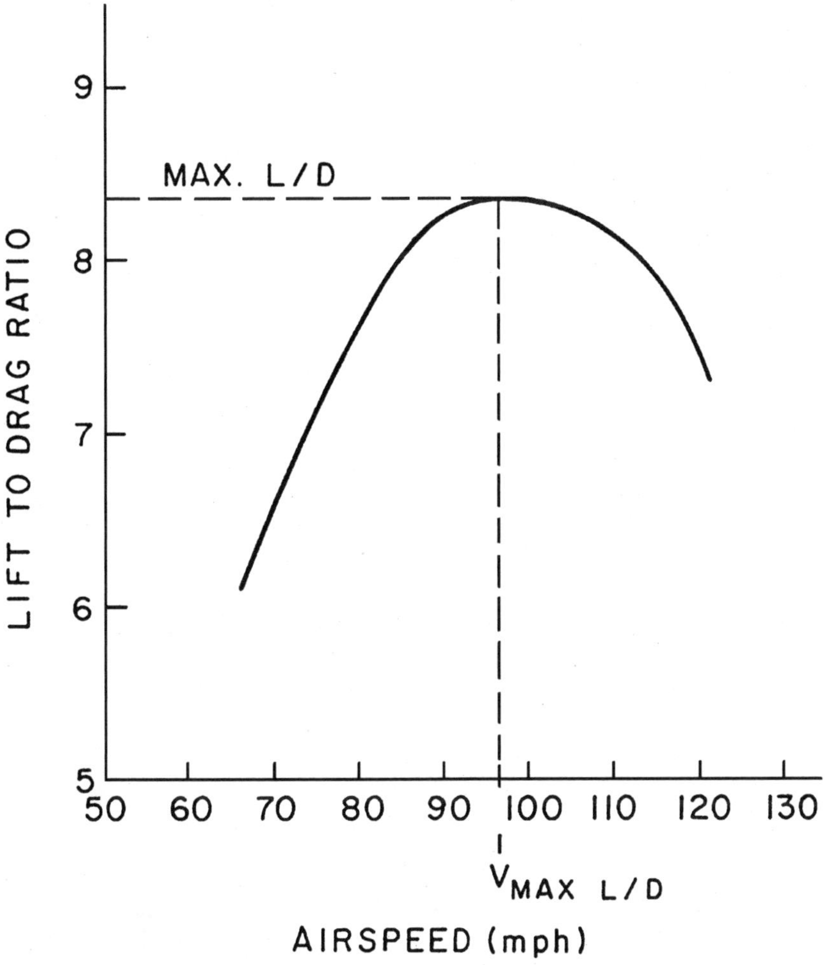

Fig. 8-3. Lift to drag ratio versus airspeed.

highest point on this curve will be your maximum L/D (or glide ratio) and the speed at this point is your best glide speed (for maximum distance). As mentioned, this speed will not change with altitude, as long as you use CAS (approximately IAS).

Weight does not affect the maximum L/D value, either, but it does affect the *speed* at which it occurs. Higher weight shifts the curve over to higher airspeeds. Since, normally, flight tests are made at less than maximum gross weight, it is desirable to correct for the best glide speed at full gross weight, sometimes referred to as "standard weight." This speed for best glide turns out to be proportional to the square root of the weight ratio.

$$V_{std} = V_{test}\sqrt{\frac{W_{std}}{W_{test}}}$$

This correction can be applied to correct your best glide speed to any weight desired. Again, it does not vary with altitude if *calibrated* airspeed is used. Using the equation given above for glide distance, the maximum distance for any altitude can be calculated by inserting your maximum L/D value:

$$d_g = (L/D)_{max} h$$

A plot of these calculations for several altitudes can be made as shown in Fig. 8-4. The glide distance could then readily be determined from this graph. Remember that you must fly at the speed for max L/D and close to the weight used in the testing.

Tests should be conducted with the engine idling. If a constant-speed propeller is involved, it should be set at the highest pitch (full decrease). This gives the minimum propeller drag. In the event of an actual engine failure, the propeller should be adjusted to this setting. If the propeller were to stop completely, and not windmill, less drag would be encountered and the glide distance would increase somewhat. Under some conditions, it might even decrease. Remember, too, that the performance predicted here is for still air. A headwind or tailwind will have considerable effect. Also, updrafts and downdrafts, which are frequent in the atmosphere, can have a very significant effort on glide performance, as glider pilots well know.

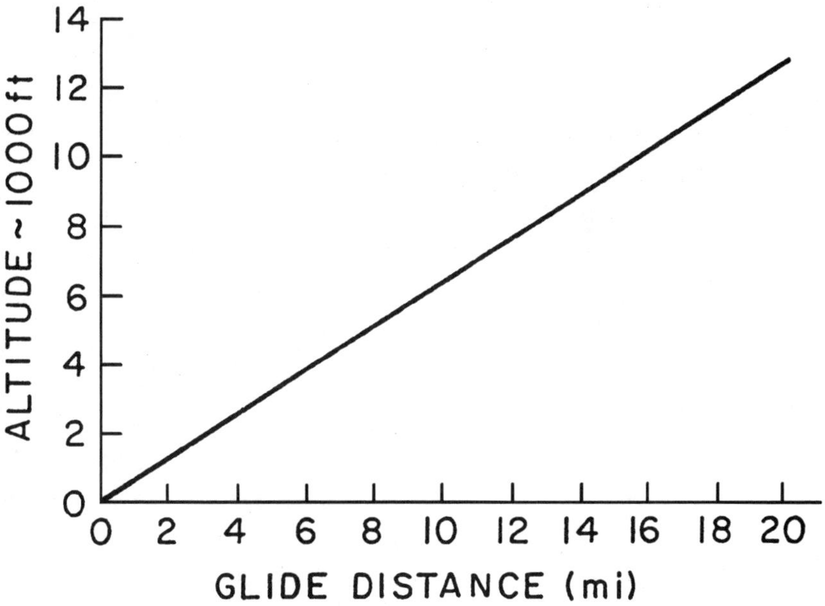

Fig. 8-4. Plot of glide distance versus altitude above ground.

For that reason, it is very important that your tests be conducted on days with very still air. Even if the air appears to be smooth, there could be subtle updrafts or downdrafts. Therefore, you should repeat the glide test for each data point (airspeed value) until the descent times at each point are fairly consistent. Even under very good conditions, there may be a fair amount of scatter in the data.

To determine glide distance in landing configuration, the entire procedure should be repeated with landing gear down and flaps as normally set for landing. This test will also yield an L/D in landing configuration which is needed to determine the air distance portion of landing, as discussed in the following section.

Landing

Landing can be thought of as the takeoff process in reverse. In takeoff, the problem was to determine the distance to accelerate from zero velocity up to takeoff velocity and then to climb over a 50-ft obstacle. In landing, you need to know the distance to glide in over a 50-ft obstacle, touch down, and then decelerate back to zero velocity. The distance is usually divided into an air distance and a ground distance, just as it was in determining takeoff performance. Some writers also consider a flare distance as a third part of the overall landing run. However, this distance can be included with the approach, or air distance, as we will do here. It should be realized that landing performance is highly dependent on pilot technique. There could be considerable variation in landing runs made by different pilots under the same conditions. A *possible* landing distance can be determined, however, just as a possible takeoff distance was considered.

Since the landing approach must be made as slowly as possible, but yet with a safe margin of speed, the stall speed plays a big role in determining landing performance. Federal Aviation Regulations require that landing distance for airplanes over 6000-lb gross weight be determined for an approach speed of not less than 30 percent above stall speed or 1.3 V_S. For lighter airplanes there is really no restriction on approach speed other than that it enables the pilot to make a safe landing. In any case, an approach speed of 1.3 V_S is a reasonable minimum approach speed. For an airplane that stalls at 60 mph, for example, this would mean a minimum approach speed of 78 mph, which is about as slow as most of us would want to fly it.

The regulations require (for those airplanes to which they apply) that the minimum approach speed be maintained down to a 50-ft height. Below that height, the speed can be further reduced. The air distance portion of the landing is thus the distance covered horizontally while the airplane is descending from 50 ft to touchdown, as shown in Fig. 8-5. If the airplane is considered to descend at a constant velocity of 1.3 V_S, then a flare distance is also required. If, on the other hand, the airplane is slowed down as the approach goes below 50 ft, the flare will be accounted for in the air distance (d_a). The more the airplane is slowed down during the air distance phase, the longer this phase will be. However, a slower touchdown speed results in a shorter groundrun phase (d_g). A full-stall landing, therefore, will result in the longest air distance but the shortest groundrun distance.

If the approach were to be made at a constant airspeed, the air distance could easily be determined by use of the glide distance

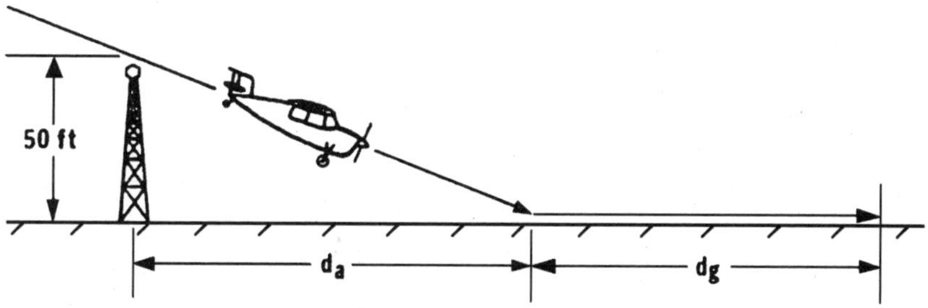

Fig. 8-5. Landing over a 50-ft obstacle showing both air and ground distances.

equation given before. In this case the altitude is 50 ft and so the equation becomes:

$$d_a = \left(\frac{L}{D}\right) 50$$

All you would have to know is your L/D ratio; this ratio multiplied by 50 is your air distance portion of landing in feet. Landing at 1.3 V_S, however, is a bit hard on the shock absorbers, to say nothing of the occupants and the rest of the airplane. A gradual flare is usually made below 50 ft of altitude. If the airplane is flared to a full-stall at touchdown and the speed is 1.3 V_S at 50 ft, then the equation for air distance is:

$$d_a = \left(\frac{L}{D}\right) (50 + 0.0305\ V_S^2) \qquad \text{(V in kn)}$$

This equation is for stall speed (V_S) in knots. The 0.0305 V_S^2 part of the equation gives the additional distance required for flaring from 1.3 V_S to V_S. If V_S is known in mph, the equation is:

$$d_a = \left(\frac{L}{D}\right) (50 + 0.023\ V_S^2) \qquad (V_S \text{ in mph})$$

Many modern airplanes can be landed (and frequently are) at a speed somewhat above stall. A touchdown speed of 15 percent above stall is often used for calculating air distance when full-stall landings are not the normal procedure. In this case, the airplane is considered to be slowed from 1.3 V_S at the 50-ft point to 1.15 V_S at touchdown. For this situation, the above equations become:

$$d_a = \left(\frac{L}{D}\right)(50 + 0.01624\ V_S^2) \qquad (V_S \text{ in kn})$$

$$d_a = \left(\frac{L}{D}\right)(50 + 0.01228\ V_S^2) \quad (V_S \text{ in mph})$$

The L/D ratio in these equations would actually be continually changing as the airplane is slowed down. An average value could be used, though, in calculating the air distance for landing. For the full-stall landing, the average velocity would be 15 percent above stall or 1.15 V_S and for landing at 1.15 V_S the average approach velocity would be 22.5 percent above stall or 1.225 V_S. The L/D should be evaluated at the appropriate one of these speeds. It should also be remembered that the L/D for use in landing distance determination must be measured in the landing configuration. Landing configurations, with landing gear and flaps down, are usually high-drag situations. Notice that the higher drag results in *lower* L/D ratios, which, in turn, results in shorter air distances for landing. This is the reason that flaps are normally used for landing. They steepen the approach by offering more drag and, hence, shorten horizontal distance during the descent. Higher flap deflections result in progressively lower L/D ratios and subsequent shorter approach distances.

Ground roll is analogous to the takeoff run, except that the airplane must be *decelerated* from touchdown speed to zero velocity. The inertial force of the airplane is the only force that tends to keep it moving, as shown in Fig. 8-6, assuming that the power has been chopped completely. The drag and frictional force of the tires on the runway tend to slow it down. These retarding forces are rather small and, without additional help, would allow a pretty long landing roll. Indeed, if downward slope were present, the airplane could easily roll on forever (with enough runway) relying only on drag and normal rolling friction to stop. Fortunately, additional help is usually supplied in the form of brakes. Brakes greatly increase the frictional force and provide most airplanes with reasonably short landing ground rolls. Larger airplanes are also sometimes equipped with thrust reversers, deceleration chutes, or other special drag-inducing devices. For most

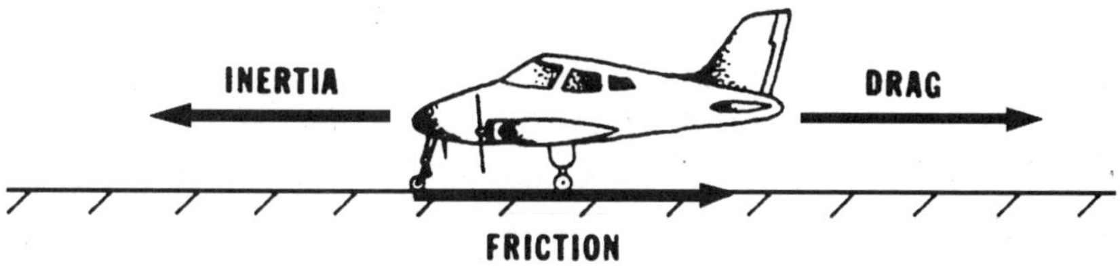

Fig. 8-6. Forces acting on airplane during landing roll.

light planes, however, the brakes are the major source of decelerating power. They are expected to be used in determining the ground distance portion of the landing.

Landing Test Procedure

The experimental part of landing performance involves the determination of the ground roll distance. As mentioned, this distance is quite variable depending on pilot technique. It depends a great deal on the exact touchdown speed, and on the amount and duration of braking during the ground roll. Even the same pilot will make different ground runs in repeated landings under the same conditions. For this reason, it is necessary to make a number of trials and to take the average of these as the resultant distance. At least three, and perhaps as many as five or six runs, will be necessary. It depends on how close the distances turn out to be on successive trials. If they are fairly close, only a few runs will be necessary. If the distances vary widely, then a larger number of trials will be required to yield a valid average figure.

The pilot should attempt to touch down at a definite speed. If he decides to obtain data for full stall landings, then the touchdown should be made in a complete stall. If landing distance is desired at a speed slightly above stall, the value of 1.15 V_S is recommended. Compute this value and attempt to touch down at that speed. For example if the stall speed in your landing configuration is 55 kn, 1.15 V_S would be 63 kn. Since the pilot will usually be sufficiently occupied in just making a smooth, precise landing, it might be wise to have an observer in the airplane monitor the airspeed indicator, and note the speed at touchdown. Remember that the landing speed of 1.15 V_S is in

TAS, so you should convert this to IAS and attempt to land at that value of airspeed, since you have to rely on the airspeed indicator to give you a reading. If the actual landing is made at a speed much different from the predetermined value, that run should be discarded and another run attempted.

During landing rollout, the pilot should apply a *moderate* amount of braking. Do not attempt to jam on the brakes immediately and with such force that the wheels skid. On the other hand, don't be too light on the brakes. Remember, that you are attempting to determine *normal* landing distance. Therefore, the brakes should be applied as in a normal landing procedure.

Spotting the actual touchdown point on the runway can be done by several observers on the ground similar to the way spot landing contests are judged. Touchdown zone stripes or fixed distance markers on the runway, as shown in Fig. 8-7, can be used as reference marks to locate the touchdown point. Runway lights could also be used, or for lack of any of these, stakes could be placed along the runway specifically for this purpose, as described in the takeoff test. Again, be sure to get the airport operator's permission to do this, and make sure that they do not pose a safety hazard to normal airport operations. Also, be sure that observers stationed along the runway are familiar with aircraft operations and remain at a safe distance from the runway. An approximate touchdown point could also be determined by an observer in the airplane using the same sort of reference markers.

Fig. 8-7. Runway markings which can be used to determine landing touchdown point.

Bring the airplane to a complete stop and, again, note the exact stopping point. This too can be done best by a ground-based observer, but is possible from the airplane. The distance, then, is determined by measuring between the touchdown point and the stopping point. This measurement could be done most accurately by use of a tape measure. However, if runway lights or other distance markers of known spacing are used, the distance could be estimated by reference to them. The points would have to by spotted as approximately 1/2 or 1/3, or whatever fraction, of the distance between them. Remember that landing distance can't be determined down to the exact foot. A length to the nearest 10 ft is probably the best that can be expected.

The tests should be conducted in calm wind, on a level runway, and at a gross weight close to the selected standard weight. Of course, the tests should also be run on the type of runway surface for which you desire the distance information. If the conditions are not exactly standard, some corrections can be made. A slight wind can be compensated for in the same way that is was in the takeoff performance. The distance measured in a slight headwind can be corrected to what it would be in zero wind by the following equation:

$$d_0 = d_w \left[\frac{V_L}{V_L - V_w} \right]^2$$

V_L in this equation is the true airspeed at landing and V_W is the headwind component of the windspeed. The term d_w is the distance as measured with a headwind equal to V_W. The resulting distance for zero wind (d_0) can then further be corrected to standard sea level by multiplying by the density ratio at the test condition:

$$d_{\text{SSL}} = d_0 \left[\frac{\rho}{\rho_0} \right]$$

If the weight at the test condition is different from standard weight, another term in the above equation would correct to standard weight. With the weight correction term added, the complete equation is:

$$d_{\text{SSL}} = d_0 \left[\frac{\rho}{\rho_0} \right] \left[\frac{W_{\text{std}}}{W_{\text{test}}} \right]^2$$

Landing ground roll distance for other altitudes and other gross weights can be calculated from the standard sea level value just as it was for takeoff ground run. The distance for higher altitudes is obtained from an equation involving the density ratio:

$$d_{alt} = \frac{d_{SSL}}{(\rho/\rho_0)_{alt}}$$

Simply divide the standard sea level value by the density ratio at the appropriate altitude. Landing roll for a lower gross weight can also be determined by first correcting the standard sea level value at standard gross weight to the new gross weight:

$$d_{new\ wt} = d_{std\ wt}\left[\frac{W_{new}}{W_{std}}\right]^2$$

The distance for this new weight can also be computed for higher altitudes by using the equation given above for altitude correction, except that you start with the distance for this new weight ($d_{new\ wt}$) in place of d_{SSL}. The resulting data can be plotted on a working chart to be used under any weight and altitude conditions as shown in Fig. 8-8.

The above discussions all involved only the ground roll distance (d_g). The total distance to clear a 50-ft obstacle would have to be

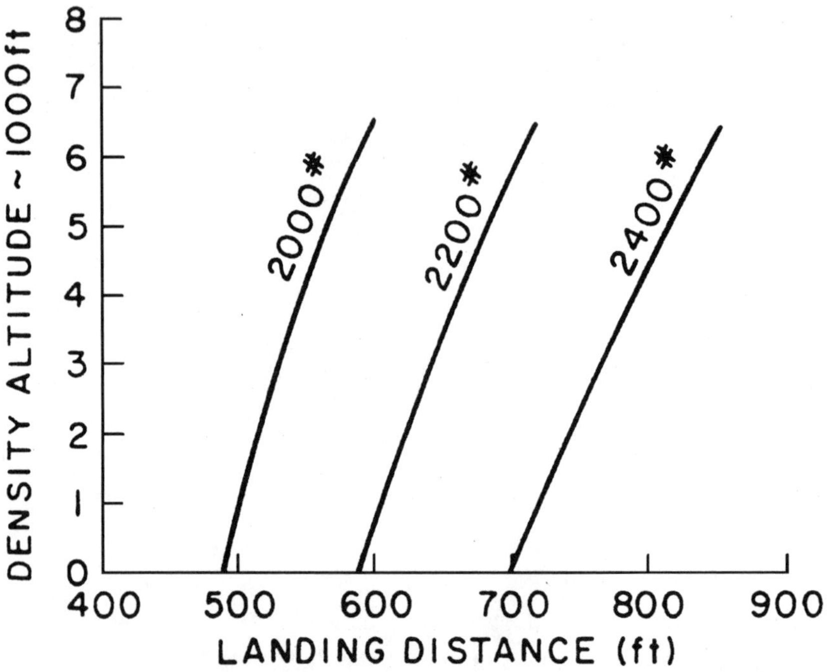

Fig. 8-8. Plot of landing distance versus altitude for various gross weights.

computed by adding the air distance to the round roll distance. The air distance would be calculated by use of the appropriate equation given in the previous section. The proper equation depends on whether a full-stall or above-stall landing is made, and whether stall speed is known in knots or miles per hour. For a full-stall touchdown and V_S in knots, for example, the air distance is calculated as:

$$d_a = \left(\frac{L}{D}\right)(50 + 0.0305\ V_S^2)$$

The L/D ratio would have to be determined experimentally as outlined in the section on glide ratio test procedure. For full-stall landing, it should be measured at 1.15 V_S.

Glide Distance Example

Test Conditions:

- Altitude change = 400 ft
- Time = 25.3 sec
- Indicated airspeed = 100 mph
- Pressure altitude = 3500 ft
- Temperature = 50°F

1. First, calculate rate of sink.

$$\text{RS} = \frac{\Delta h}{t} = \frac{400}{25.3} = 15.81 \text{ ft/sec}$$

2. Calculate true airspeed.
   - Assume, from airspeed calibration curve: 100 mph IAS = 102 mph CAS
   - From calculator (E6B or electronic): 102 mph @ 3500 ft and 10°C = 108 mph TAS
3. Now, determine L/D ratio.

$$L/D = \frac{1.467\ V}{\text{RS}} = \frac{1.467(108)}{15.81} = 10.02$$

4. Plot this point against indicated velocity as shown in Fig. 8-9.

5. Assuming similar calculations for other velocities have yielded the curve as shown, determine $(L/D)_{max}$ as the highest point on this curve.

$$(L/D)_{max} = 10.4$$

6. The glide distance can now be determined for any altitude. For 10,000 ft the distance is determined as:

$$d_g = \frac{(L/D)_{max}h}{5280} = \frac{10.4(10{,}000)}{5280} = 19.7 \text{ mi}$$

Landing Distance Example

Test Conditions:

- Average ground roll = 600 ft

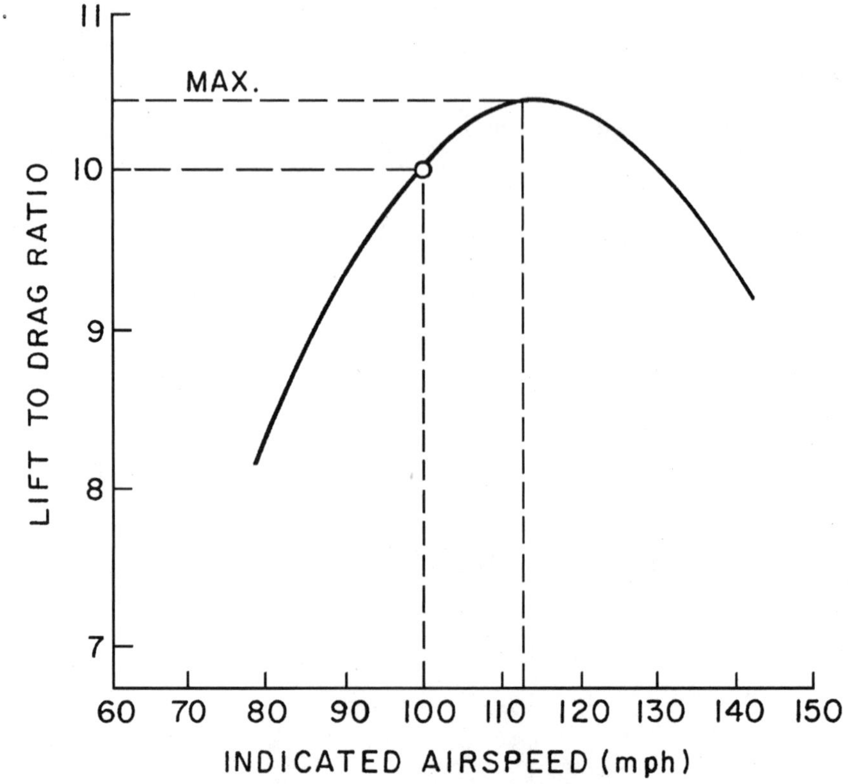

Fig. 8-9. Determining maximum *L/D* ratio from flight tests.

- Headwind = 5 kn
- Pressure altitude = 2000 ft
- Temperature = 38°F
- Weight = standard weight
- Stall speed = 55 kn

Assume that the landing is to be made at full stall.

1. First, correct the ground roll to zero wind conditions.

$$d_0 = d_w \left[\frac{V_L}{V_L - V_w} \right]^2 = 600 \left[\frac{55}{55 - 5} \right]^2 = 726 \text{ ft}$$

2. Next, correct this value to standard sea level distance. From Figs. 1-3 and 1-4, a pressure altitude of 2000 ft and temperature of 38° yield a value of ρ/ρ_0 of 0.97.

$$d_{SSL} = d_0 \left[\frac{\rho}{\rho_0} \right] = 726\ (.97) = 704 \text{ ft}$$

This is the ground roll distance at standard sea level, now referred to as d_g.

3. Now calculate the air distance to clear a 50-ft obstacle. For full stall landing, the average approach speed is 1.15 V_S.

$$1.15\ V_S = 1.15\ (55) = 63 \text{ kn}$$

From a plot of L/D versus velocity in landing configuration, assume that we have obtained an L/D of 6.9 for $V = 63$ kn. From the equation for air distance for full stall landing and stall speed in knots, we can now obtain the air distance,

$$d_a = \left(\frac{L}{D} \right) (50 + 0.0305\ V_S^2) = (6.9)\ [50 + 0.0305\ (55)^2] = 981 \text{ ft}$$

4. The total landing distance to clear a 50-ft obstacle is now:

$$d = d_g + d_a = 704 + 981 = 1685 \text{ ft}$$

CHAPTER 9

Computer Methods

Computers are regarded by most people as tools for accessing information or quickly sending mail. They were originally conceived, however, to compute complex mathematical operations. This application is even more important today in the scientific community. Fortunately, the solution of many computational problems can be done rather simply with modern programs, without the user really needing to understand the process. One of the most convenient of such programs is known as a spreadsheet program, and one of the most popular of these is Microsoft Excel®. This software will, therefore, be employed in this chapter. It was written using Excel 97, although version 2000 has since come on the market. This latest, or earlier versions, could also be used with only some very minor differences.

The programs will include all of the experiments covered in previous chapters, so that it gives you an alternate method of reducing the data. One of the advantages of a computer program is that it can calculate results much more accurately than some of the approximate methods described previously. More precise mathematical equations can be incorporated, but the user does not necessarily need to understand them. Inserting numbers into an equation is not only more accurate, but also more convenient than going to a graph, and estimating a value from the curve. Excel can also plot curves in a very quick and simple fashion to display the trend in the performance item being considered, and to provide working charts similar to those in the standard pilot operating handbooks.

One such application frequently used throughout the book is calculating air density. This operation is greatly simplified by using the exact mathematical relationship that temperature and pressure altitude have on density, as described in Chap. 1. It will be incorporated in all of the programs in which it has a role here, but, of course, the computer will take care of it automatically. All the user has to do is to insert the appropriate temperature and pressure altitude. The same is true of determining density altitude.

All tests are assumed to have been run by the methods described in the previous chapters. The data, however, will be entered into the spreadsheet, and the program will then calculate the results. A sample spreadsheet is given for each test. The user simply has to substitute his own data for that shown in the sample. All mathematical equations used are listed here, in case you do not have access to the software supplied. In cases where formulas are copied to succeeding cells in a column, only the formula for the first row of the column is given. The formulas are automatically corrected in succeeding cells by the copy process. The graphs of most of the performance data resulting are also presented. It is assumed that the user is familiar with the basic operation of Excel, both in constructing spreadsheets and in plotting graphs.

Note that the example spreadsheets and graphs in the figures in Chap. 9 are not all for the same airplane. Do not try to apply data from one example to another. Also, note that, in some examples, the data is plotted in airspeed values of miles per hour, while in others it is in knots. Many older airplanes and homebuilts will have indicators in miles per hour, while newer craft are all in knots, so it's simpler to use whatever values you read. It doesn't matter which values you use; just be consistent. The conversion from one value to the other is pretty easy. Knots are miles per hour times 1.15, and miles per hour is knots divided by that number.

Graphs are easily constructed by use of the Excel Chart Wizard. All graphs should be plotted with the XY (Scatter) option. After plotting the data, fit a trendline to them. Some charts will require a Polynomial curve fit (order 2), while others use the Linear trendline (the default option). In some cases the equation of the resulting curve is required, an option under the Trendline selection. Make sure that the appropriate items are on the appropriate axes, X or Y, when fitting an equation to the curve. If they are reversed, correct them by selecting Source Data from the Chart toolbar, then select Series, and make the changes in the appropriate boxes. You may or may not want to include the legend on your charts. In the case of multiple curves, it may be helpful, as depicted on the rate of climb curves in Fig. 9-10. An alternative would be to label the curves, as shown on the takeoff chart in Fig. 9-8. This method might be more desirable in your final working charts.

A floppy disk is included with the book containing all of the sample spreadsheets and charts referred to in this chapter. Insert it into the floppy drive, and select it from the Excel files menu. The menu for the floppy will then display a file for each of the following performance items, which may be selected and opened as required. Since each file contains data for a sample test, you may want to make a copy to work from, so that you do not destroy the data on the sample. You will have to enter your own data in place of those on the sample.

Airspeed Calibration

Figure 9-1 is a sample spreadsheet for airspeed calibration. Since there are several ways described in Chap. 3 to actually conduct this test, it is assumed that groundspeed in each direction is available for input to the program. These data would be obtained directly from the GPS method, but would have to be calculated by dividing time into distance for the speed-course method.

Into column A, insert the groundspeeds in one direction (Vg1), and in column B list the groundspeed in the other (Vg2). The pressure altitude must be inserted into cell C4, and temperature (in Fahrenheit) into D4. The program then calculates the density ratio in E4. In case you do not have the software, the formula for this calculation is

	A	B	C	D	E	F
1			Airspeed Calibration			
2						
3			P Alt (ft)	Temp (F)	Dens Rat	
4			3300	58	0.887846	
5						
6	Vg1(kn)	Vg2(kn)	Vgav (Vt)	Vi (kn)	Vcal (kn)	
7	124	117	120.5	115	113.5418	
8	120	109	114.5	110	107.8883	
9	116	106	111	105	104.5904	
10	112	99.2	105.6	100	99.50223	
11	103	97.5	100.25	95	94.46116	
12	96	89.5	92.75	90	87.39424	
13	86.5	80.5	83.5	80	78.67838	
14	80	75	77.5	75	73.02484	
15	75	71	73	70	68.78469	
16						
17						
18						
19						
20						

Fig. 9-1. Spreadsheet for airspeed calibration.

= ((1 − 0.00000689*$C$4) ^ 5.256)*519/($D4 + 460)

Column C calculates the average groundspeed (Vgav), which is the true airspeed. These values are simply the values in column A plus those in column B divided by 2. Calibrated airspeed (Vcal) is calculated in column E. This operation involves the density ratio calculated in E4. The formula is

= C7*SQRT($E$4)

Note that, if your groundspeed is entered in knots, the resulting Vcal will also be in knots. If you want to convert these values into miles per hour, simply multiply them by 1.15.

Enter into column D the corresponding indicated values (Vi) that you flew while measuring the groundspeeds. Again, you want to be consistent in using either knots or miles per hour. The example shown is for a Cherokee. Now, plot calibrated airspeed versus indicated airspeed as shown in the chart for this example in Fig. 9-2, and you have your airspeed calibration curve. Fit a trendline to these data points, and have the program print out the equation of the curve. Normally, the data will be in nearly a straight line, so that a linear fit is appropriate. If the data appear to be on more of a curve, you may need a polynomial fit. In the example, a linear fit is made, and the equation is:

$$y = 1.0075x - 2.0525$$

The x is the Vi value, and the y represents the resulting value of Vcal. This equation can then be used for correcting your indicated airspeed to calibrated for all subsequent tests. In the remaining programs, the numbers from this equation are entered in cells labeled "Vcal factors." The first number (multiplier of the x) is entered into the first cell, and the second number into the adjacent cell. Be sure to include the minus sign if it exists, as is the case here.

Stall Speed

Stall speeds are measured according to the procedures outlined in Chap. 4. They should be done for at least clean and full-flap landing configurations. Speeds for intermediate flap settings could also be done, at the user's discretion. The spreadsheet, as shown in Fig. 9-3, is for a Cherokee, and includes all four discrete flap settings, listed in cells C7 through C10. The respective stall speeds are entered into B7 through B10. The takeoff gross weight (in pounds) is entered into C4, and the fuel consumed (gallons) for the entire flight in D4. Cell E4 then calculates the average weight during the tests. The formula for this determination is

= ($C$4 + ($C$4 – 6*$D$4))/2

The stall speed for any other gross weight can then be calculated by using a ratio of the desired weight to the average test weight. The various weights chosen in this case are listed in column A, starting in row 13. Column B then calculates the respective stall speeds for 0 degrees flaps, and column C does the same for 40 degrees flaps. The formula for the first weight and zero flaps is

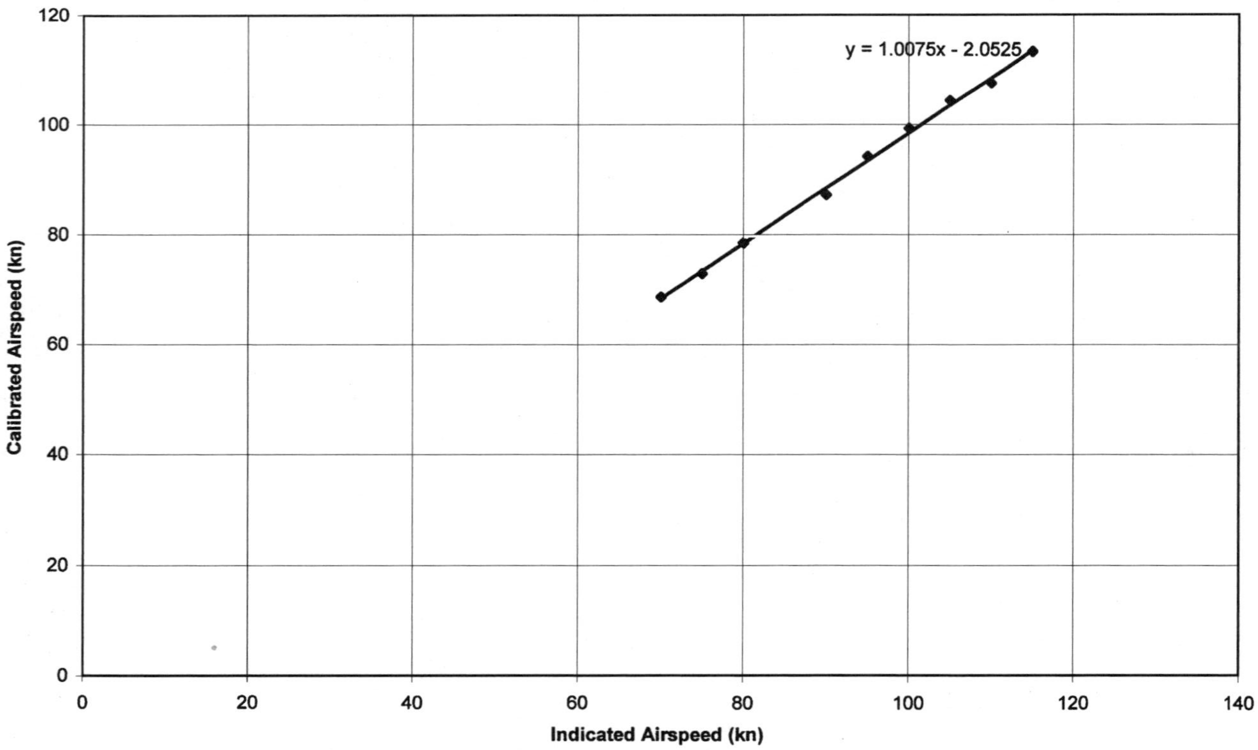

Fig. 9-2. Plot of airspeed calibration curve.

	A	B	C	D	E
1			**Stall Speed**		
2					
3			**W to**	**Fuel Burn**	**W av**
4			2400	12	2364
5					
6		**V ind**	**Flaps**		
7		66	0		
8		63	10		
9		60	25		
10		56	40		
11					
12	**W**	**V (0)**	**V (40)**		
13	1600	54.29754	46.07064		
14	1800	57.59124	48.8653		
15	2000	60.7065	51.50855		
16	2200	63.66951	54.02262		
17	2400	66.50064	56.42478		

Fig. 9-3. Spreadsheet for stall speed determination.

= $B$7*SQRT(A13/$E$4)

For 40 degrees, $B$10 is substituted for $B$7. It doesn't matter if the speeds are initially entered in knots or miles per hour; the resulting speeds will be in the same units. The example is done in miles per hour, since that is the way the airspeed indicator on the test aircraft was calibrated. A graph would be helpful to the pilot to show the stall speed for the particular weight at a given flight condition. A plot of columns B and C against column A, beginning at row 12 is shown in Fig. 9-4. The speed is shown in indicated airspeed because that is the most useful to the pilot. Make a polynomial fit to each curve.

Takeoff Performance

Takeoff performance is a bit more complicated to compute; hence the Excel program is quite helpful here. One of the major problems

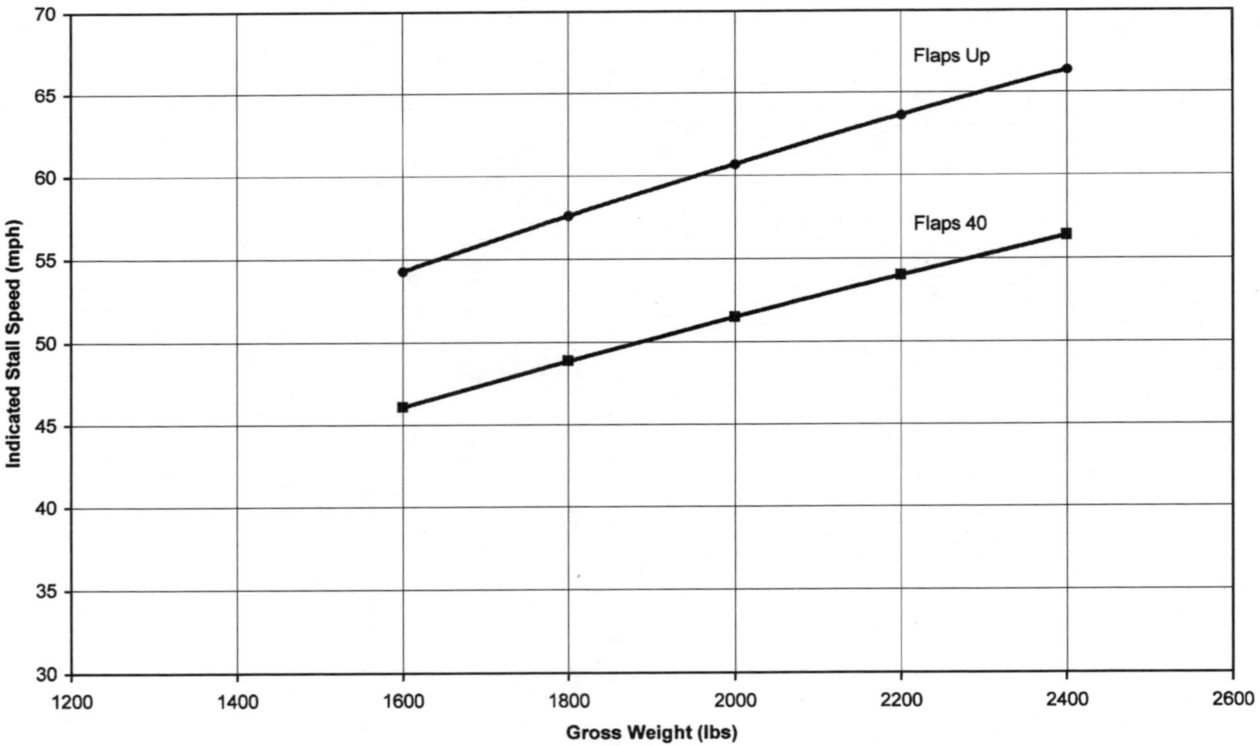

Fig. 9-4. Plots of stall speed versus gross weight for two flap settings.

is the difficulty in measuring the exact time for reaching a given velocity during the takeoff run. To solve this problem, we *calculate* the velocity from the data of distance versus velocity. This approach involves a calculus procedure known as "differentiation." It can be accomplished, however, rather simply, and without having even a single lesson in calculus.

Using the distance-versus-time data obtained by one of the methods described in Chap. 5, list these values in columns A and B under t and d of the sheet shown in Fig. 9-5. Then plot these data on a chart similar to Fig. 9-6, making sure that distance is on the vertical axis. Do a polynomial curve fit (to the order of 2), and also have the equation of the curve printed on the chart. This equation describes, mathematically, the relationship of any distance to the time it took to reach it. The equation for the example is:

$$y = 2.3736x^2 + 9.5083x - 13.513$$

Of course, the y in this case is distance, and the x is time. We could write the general form of this equation as

$$y = Ax^2 + Bx + C$$

where A, B, and C are the numbers in the equation.

	A	B	C	D	E	F	G	H	I
1				Takeoff					
2									
3	Wstd (lbs)	W (lbs)	Press Alt	OAT (F)	Hdwind	Vcal factors			
4	2400	2200	2000	38	5	1.007	*-2.05*		
5									
6	t (sec)	d (ft)	V (ft/sec)	V (kn)		Vsi (kn)	Vst (kn)	Vto (kn)	
7	0	0	0	0		47	46.00072	55.20087	
8	5.5	100	35.61818	21.07584					
9	8	200	47.4863	28.0984					
10	10	300	56.9808	33.71645		D Alt	dstdw	d2000	
11	11.5	400	64.10168	37.92999		0	1189.889	826.3121	
12	12.7	500	69.79838	41.30081		1000	1276.959	886.7768	
13	14	600	75.9698	44.95254		2000	1371.073	952.134	
14	15.3	700	82.14123	48.60428		3000	1472.858	1022.818	
15	16.5	800	87.83793	51.9751		4000	1582.999	1099.305	
16	18.5	1000	97.33243	57.59315		5000	1702.249	1182.118	
17	21	1200	109.2006	64.61571		6000	1831.436	1271.83	
18									
19									
20									
21			d(to) =	893.5454	d(0wind)=	1065.126	d(stdsl)=	999.8376	
22			d(stdw)=	1189.889					

Fig. 9-5. Spreadsheet for takeoff distance calculation.

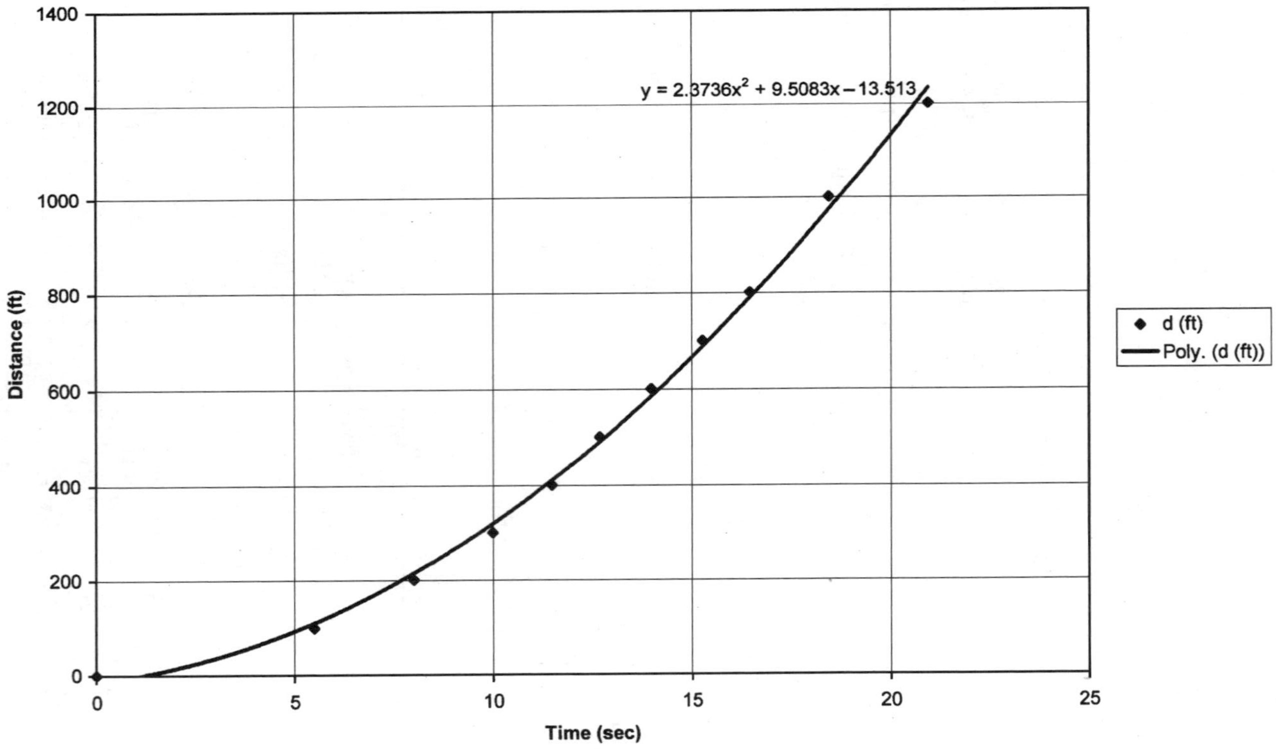

Fig. 9-6. Plot of distance versus time on takeoff run.

Now, velocity is really what is known as the derivative of distance with respect to time. It represents the slope of the time-distance curve at each minuscule point along the curve. To calculate it we multiply 2 times A, multiply that times x, and then add B. This equation is

$$y = 2Ax + B$$

Or, in this case,

$$y = 4.7472x + 9.5083$$

This is the equation for velocity. You will have to insert the proper numbers for your takeoff distance curve into cell C8, and copy it to the rest of the column. In Excel language, this formula becomes

= 4.7472*A8 + 9.5083

We now have the velocities associated with each time and distance point along our takeoff run. Column D changes the velocities into knots. Plot these values versus distance (not time) in another curve, as shown in Fig. 9-7, and do a polynomial curve fit. Place the velocity on the vertical, or y, axis. All we need to do now is to enter this curve with the defined takeoff velocity (20% above stall speed is used here), and read the corresponding takeoff distance. First, though, we must enter the appropriate stall speed for the weight in the test (from your curve similar to Fig. 9-4) in F7. In Fig. 9-4, this value would be about 63 mph for the flaps-up condition, or 55 kn. For this example, let's assume that it is 47 kn. Also enter the Vcal factors in F4 and G4. The program then calculates true airspeed for stall in G7, and takeoff speed in H7. This is the value we use to enter the curve.

We could have the computer calculate this value by again having it plot the equation of the curve in Fig. 9-7, and entering it into cell D21. You will have to enter the numbers resulting on your curve, but in the example, the formula is

= 0.3084*H7 ^ 2 − 0.6334*H7 − 11.227

The program then lists the takeoff distance for the test conditions as 893.5454 ft.

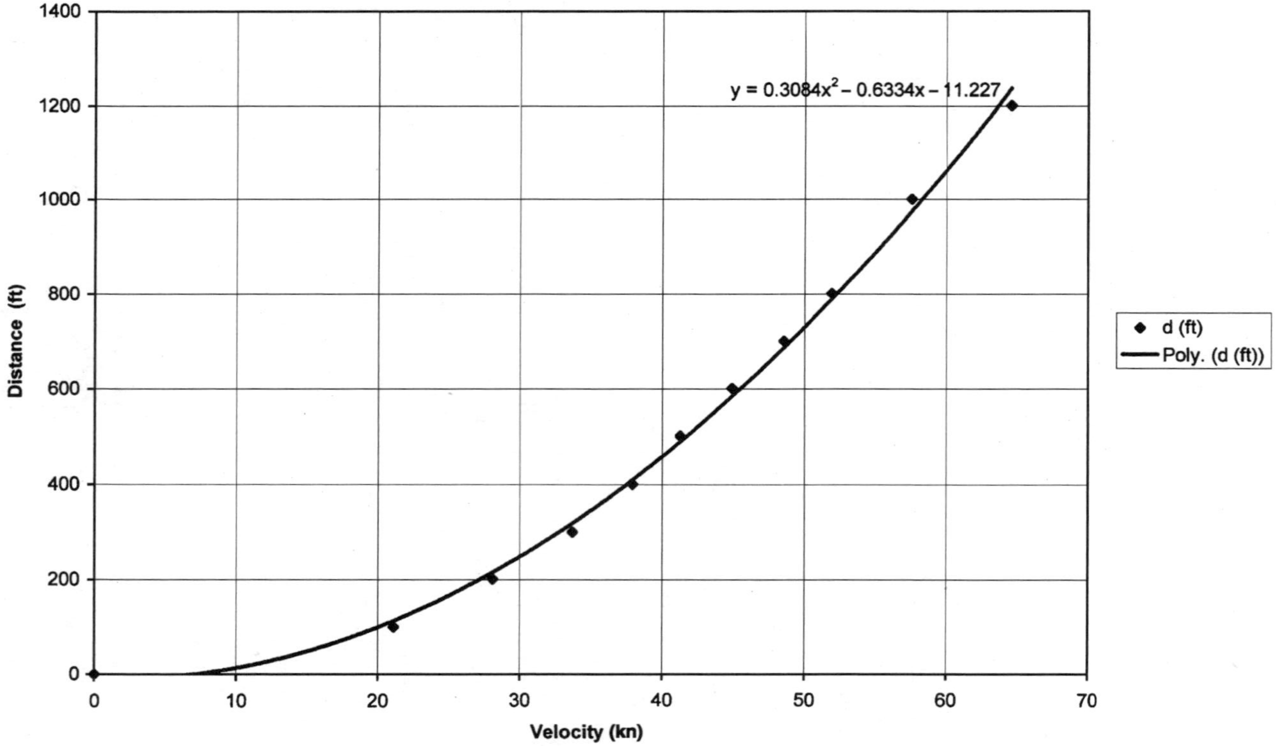

Fig. 9-7. Plot of velocity versus distance on takeoff run.

This distance is for the airplane in the test conditions. If the test was not run with zero wind, or at standard weight and sea level altitude, it is desirable to correct to those conditions for a baseline takeoff distance. Enter the takeoff weight into cell B4. It is assumed that the tests will be done close enough to this weight so that weight averaging is not required here. Then, put pressure altitude in C4, outside air temperature in D4, and the measured headwind in E4.

The wind correction is then done in cell F21. This formula is

= ((H7/(H7 − E4)) ^ 1.85)*D21

This is essentially the same equation that was used in Chap. 5, except that the ratio of takeoff velocity to takeoff velocity minus wind was squared there for easier handling by a simple calculator. The exponent of 1.85 gives a more accurate result, and is used here, since it can easily be handled by the computer.

Now we must correct to standard sea level atmospheric conditions. This function is performed in cell H21, and uses the formula:

= F21*(((1 − (0.00000689*C4)) ^ 5.256)*519/(D4 + 460)) ^ 2

This value is then finally corrected for standard weight, which should be entered initially in cell A4. The final takeoff distance is then calculated in D22, using the formula:

= H21*(A4/B4) ^ 2

This value is the distance for takeoff at zero wind, standard sea level, and standard weight (2400 lb). For the example, the value is 1190 ft (rounded off).

You can now calculate takeoff distance for other weights and altitudes. Column F lists the various density altitudes up to 6000 ft in rows 11 to 17. In cell G11, the sea level value for 2400 lb is reproduced from D22. The value for this weight at 1000 ft altitude is then determined in G12 with the formula:

= $G$11/((1 − 0.00000689*F12) ^ 4.256) ^ 2.4

This is then copied to the succeeding cells in that column, and gives the takeoff distance for the respective altitudes. Cell H11 gives the takeoff distance at sea level for 2000 lb. This is done by use of the formula:

= G7*(2000/A4) ^ 2

The values for higher altitudes at that gross weight then are given in the following cells in that row, by the formula:

= $H$11/((1 − 0.00000689*F12) ^ 4.256) ^ 2.4

You could calculate values for any other gross weight desired by simply entering the new gross weight into the formula for cell G11 in place of the 2000 value. The chart for the 2400 and 2000 lb weight is shown in Fig. 9-8. Use a polynomial fit to construct a curve to the data points.

Rate of Climb

Figure 9-9 displays the spreadsheet for rate-of-climb calculation. Data should be taken according to the procedures in Chap. 6. The takeoff weight and fuel burn in gallons should be listed in D4 and E4 respectively. The program then calculates the average weight during the tests, the starting weight plus the ending weight divided by 2. Ending weight is takeoff weight minus 6 times gallons of fuel consumed (6 lb/gal). This average weight is then used for the calculations. If portions of the tests are performed at different times when weights are significantly different, separate sheets should be used.

In column A, list the appropriate pressure altitude, and in column B, the OAT in degrees Fahrenheit. Place in column C the altitude increment through which you climbed, in column D the time it took in seconds, and in column H the indicated airspeed during the climb. Usually, we attempt to fly even speed increments, such as every 5 kn, but if the speed averaged a little more or less than attempted during the climb, list the exact value (81, 86, 95, etc, in the example). These three values, altitude increment, time, and airspeed, are the primary data entered.

The program then calculates the rate of climb. First the altitude increment measured (Del Hm) is corrected for nonstandard temperature to actual altitude increment (Del Ha) by the formula:

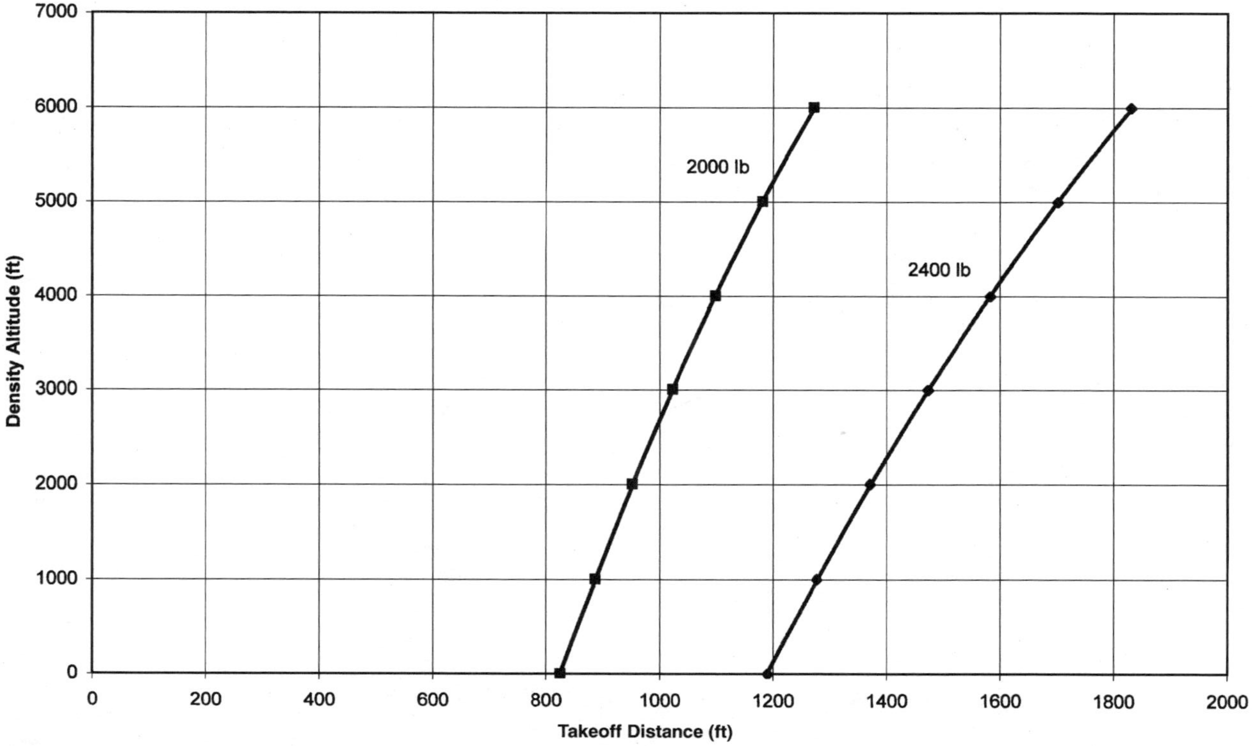

Fig. 9-8. Plots of takeoff distance versus altitude for two gross weights.

	A	B	C	D	E	F	G	H	I
1					Rate of Climb				
2		Gear Up	Flaps Up						
3				W to	Fuel Burn	W av			
4				2610	21.7	2544.9			
5									
6	P Alt	OAT	Del Hm	Time	Del Ha	Dens Rat	D Alt	V ind	R/C
7	(ft)	(Deg F)	(ft)	(sec)	(ft)		(ft)	(kn)	(ft/min)
8									
9	2000	36	400	29	387.6149	0.972774	938.4327	81	801.9619
10	2000	36	400	27.4	387.6149	0.972774	938.4327	86	848.7918
11	2000	36	400	26	387.6149	0.972774	938.4327	95	894.496
12	2000	36	400	27.4	387.6149	0.972774	938.4327	104	848.7918
13	2000	36	400	29.7	387.6149	0.972774	938.4327	108	783.0605
14	2000	36	400	38.2	387.6149	0.972774	938.4327	115	608.8193
15	2000	36	400	52	387.6149	0.972774	938.4327	120	447.248
16	4000	45	400	41.4	400.2407	0.887325	4020.578	75	580.0589
17	4000	45	400	35.5	400.2407	0.887325	4020.578	81	676.4631
18	4000	45	400	33.8	400.2407	0.887325	4020.578	87	710.4864
19	4000	45	400	32.7	400.2407	0.887325	4020.578	96	734.3865
20	4000	45	400	36.4	400.2407	0.887325	4020.578	103	659.7373
21	4000	45	400	55	400.2407	0.887325	4020.578	109	436.6262

Fig. 9-9. Spreadsheet for rate-of-climb calculation.

$$= ((B9 + 460)*C9/(519*(1 - 0.00000689*A9))$$

and listed in column E. The density ratio and density altitude are calculated by the standard formulas, which for this spreadsheet are

$$= ((1 - 0.00000689*A9)\text{^}5.256)/((B9 + 460)/519)$$

$$= (1 - F9\text{^}\ 0.235)/0.00000689$$

and these are listed in columns F and G, respectively. The values are given for the first row (9), and the program copies them into succeeding rows.

The rate of climb in feet per minute is then calculated in column I by

$$= (E9/D9)*60$$

Note that the two density altitudes resulting in the example are 938 and 4020 ft. The data for each of these altitudes are then plotted on Fig. 9-10. Make a polynomial fit to each of these curves (order 2), and then note the maximum value for each of the altitudes. Also read the airspeed value corresponding to the maximum rate-of-climb point, which would be your best rate-of-climb speed, or V_y. These curves are

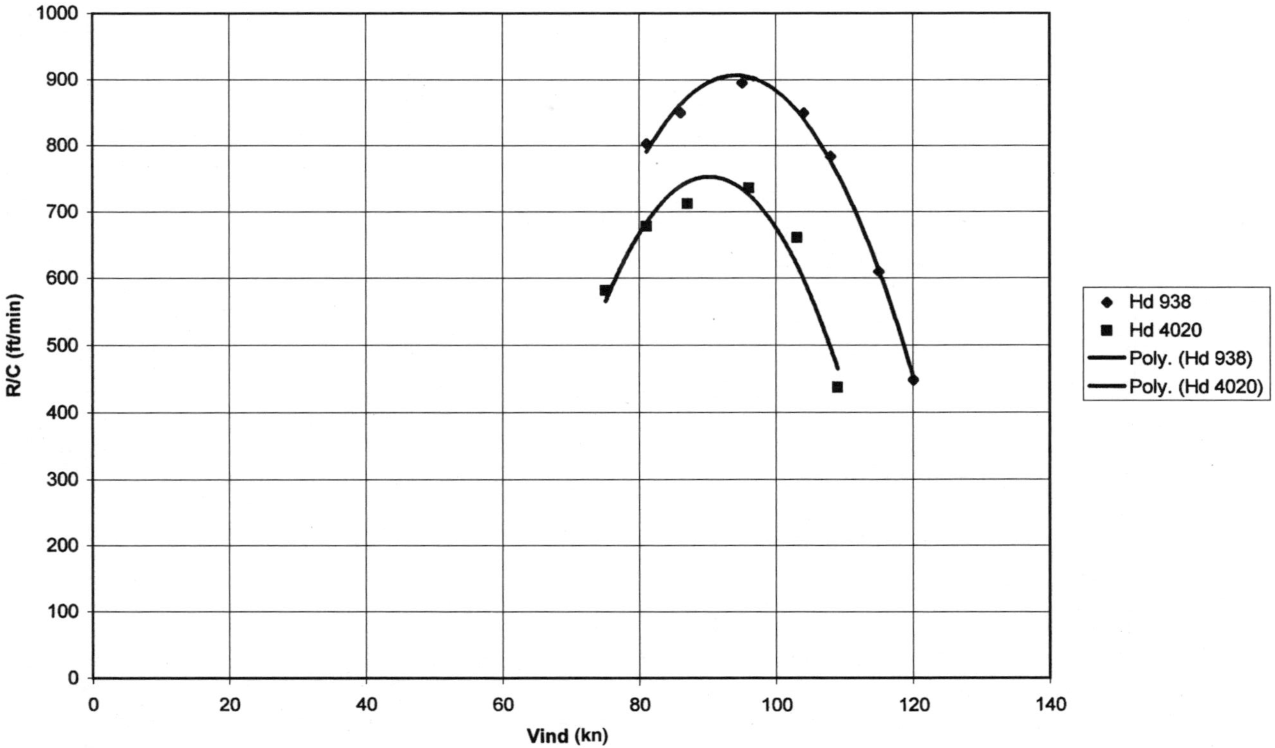

Fig. 9-10. Plot of rate of climb versus airspeed for two altitudes.

plotted for indicated airspeed rather than true, since that is what the pilot will read. This is entirely optional, however.

The spreadsheet in Fig. 9-11 shows the maximum R/C for the two altitudes plotted, and for three other altitudes: 7600, 11,100, and 13,100 ft. (Note that R/C is used here as the symbol for rate of climb; it is the same as the sometimes used RC, and does not imply a ratio.) This sheet then corrects the R/C to that for standard weight (or any other weight) as entered into cell E3. This calculation is a bit complicated, but can be done easily by the program. It requires, however, the wingspan and the wing efficiency factor (e), which is input into F3 and G3, respectively. The e value is hard to determine accurately, but can be estimated pretty well as 0.6 for low wing airplanes, and 0.8 for high wings, since it is affected by the fuselage interference on the wing.

True airspeed is also required, so enter the appropriate Vcal factors into cells C3 and D3. For this example, no airspeed calibration was available, so it was assumed that the Vcal was equal to the Vi. Hence, the 1 and 0 factors entered. In any of the tests, if your airspeed is found to be highly accurate (no appreciable error between Vi and Vcal), you can simply enter a 1 and 0 to make these airspeed values the same. Input the indicated value(s) of Vy in column F, and the program then converts this to true airspeed in column G. In the example, a constant value of 95 kn is used for Vy indicated.

Density is also needed, so it is calculated in column E from the density altitude in column B, using the formula:

	A	B	C	D	E	F	G	H	I
1				R/C versus Altitude					
2			Vcal factors		Ws (lbs)	Span (ft)	e (.6 lo,.8 hi)		
3			1	0	2700	35.4	0.6		
4									
5	R/C max	Dens Alt	W	Del W	Density	Vy ind	Vy true	Del R/C	R/C stdw
6	(ft/min)	(ft)	(lb)	(lb)	(slug/cuft)	(kn)	(ft/sec)	(ft/min)	(ft/min)
7	905	938	2544.9	-155.1	0.002312	95	162.7805	-97.0221	807.9779
8	750	4020	2544.9	-155.1	0.002109	95	170.4391	-89.5454	660.4546
9	550	7600	2544.9	-155.1	0.001891	95	180.0185	-79.8201	470.1799
10	370	11100	2544.9	-155.1	0.001694	95	190.1689	-71.4605	298.5395
11	300	13100	2544.9	-155.1	0.001589	95	196.3511	-68.7844	231.2156
12		17510							0
13		0							851

Fig. 9-11. Spreadsheet for determining standard weight rate of climb at various altitudes.

= 0.002377*((1 − B7*0.00000689) ^ 4.256)

and the difference in the new weight from the original weight listed in column C, determined in column D as Del W. The increment of R/C for the new weight can then be calculated in column H as

= D7*((A7/C7) + 240/(3.1416*E7*$G$3*($F$3) ^ 2*G7))

This change in R/C is called Del R/C, and is added or subtracted (depending on whether it is + or −) to the original R/C value from column A, for the final R/C in column I (R/C stdw). Usually, Del R/C will be negative, since the tests are normally run at a lower weight than standard. You could calculate the R/C for any other weight by simply inputting the appropriate new weight in cell E3.

A plot of the R/C versus density altitude is made in Fig. 9-12. Make a linear curve fit to these data, and have the program print out the equation of the line. The second number in the equation is the absolute altitude. The sea level rate of climb can be determined by dividing the second number by the first (the coefficient of x), but make it a positive value. In the example,

$$y = -20.577x + 17{,}510$$

The absolute ceiling, or altitude where R/C is 0, is 17,510 feet, and the sea level rate of climb is

$$\frac{17{,}510}{20.577} = 851$$

These data points can then be added to the sheet in Fig, 9-11, and plotted to yield the complete R/C versus altitude curve shown in Fig. 9-13.

Cruise Performance

Cruise performance includes both cruising speed and range. Figure 9-14 shows the spreadsheet for these calculations. Data should be collected according to procedures in Chap. 7. Enter the appropriate Vcal factors into cells B4 and C4. While not required for the calculations, the weight should be known, so enter the takeoff weight into D4, and

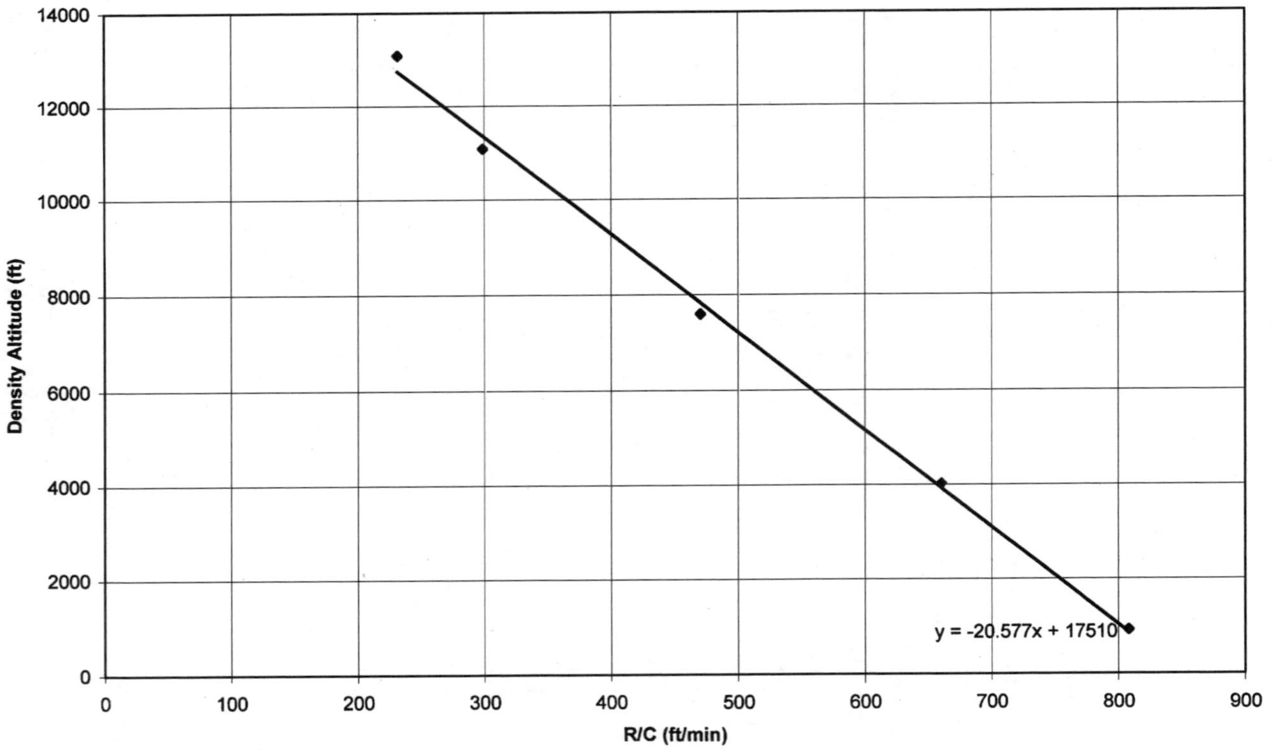

Fig. 9-12. **Plot of standard rate of climb at various altitudes.**

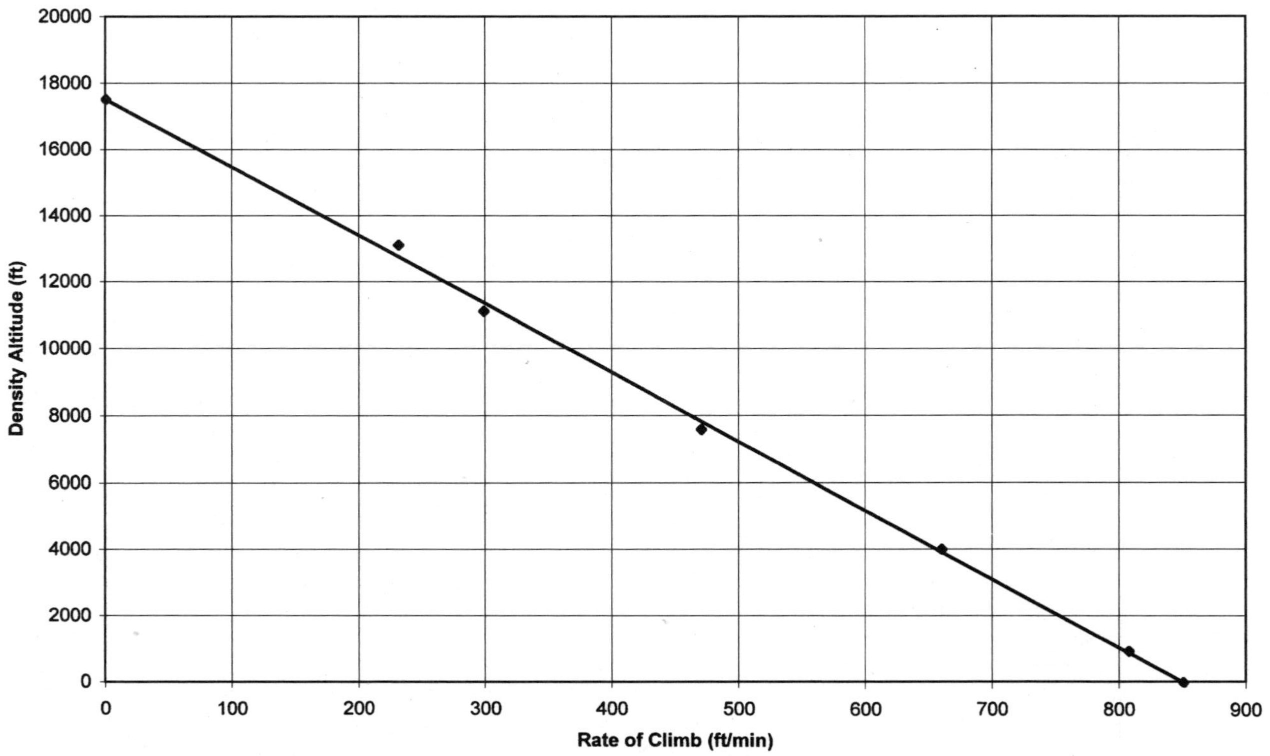

Fig. 9-13. Extrapolated plot of rate of climb versus altitude for ceiling determination.

	A	B	C	D	E	F	G	H	I
1				Cruise Speed and Range					
2									
3		Vcal factors		Wto (lb)	Fuel Burn	Fuel Cap			
4		0.9636	3.6548	2700	12.5	72			
5									
6		Vi (kn)	Press Alt	OAT (F)	FF(gal/hr)	Dens Rat	Vtrue (kn)	Rang(nm)	Dens Alt
7	65%	145	2000	52	10	0.942375	147.6954	1063.407	2010.293
8		144	3000	48	10	0.915433	148.8459	1071.69	2982.576
9		142	4000	45	10	0.887325	149.1391	1073.802	4020.578
10		141	5000	42	10	0.859884	150.461	1083.319	5058.525
11		139	6000	38	10	0.834771	150.5982	1084.307	6030.854
12		138	7000	34	10	0.810223	151.792	1092.902	7003.154
13		137	8000	30	10	0.786232	153.0037	1101.627	7975.423
14		135	9000	27	10	0.761223	153.2879	1103.673	9013.43
15		134	10000	23	10	0.738354	154.5223	1112.56	9985.708
16									
17	75%	151	1000	55	11	0.971803	151.3069	990.3721	972.2752
18		149.5	2000	51	11	0.944219	152.0136	994.9978	1944.521
19		147.5	3000	48	11	0.915433	152.3708	997.3363	2982.576
20		146	4000	45	11	0.887325	153.2309	1002.966	4020.578
21		145	5000	41	11	0.8616	154.4635	1011.034	4992.869
22		143	6000	38	11	0.834771	154.8168	1013.346	6030.854
23									

Fig. 9-14. Spreadsheet for cruise speed and range calculation.

the fuel burn in gallons into E4. Fuel capacity is necessary, and entered into F4.

This sheet is set up for measurements at 65 and 75 percent power. If you want any other settings, you could include them below, and copy the equations from the last row in columns F through I down through as many rows as you need. Enter your data for indicated airspeed, pressure altitude, outside air temperature, and fuel flow into columns B, C, D, and E, respectively.

The program will calculate density ratio in column F from the formula:

$$= ((1 - (0.00000689*C7)) \wedge 5.256)*519/(D7 + 460)$$

Using this ratio, the density altitude is calculated by

$$= (1 - (F7 \wedge 0.235))/0.00000689$$

and listed in column I. The cruising speed in true airspeed is calculated in column G from

$$= (\$B\$4*B7 + \$C\$4)(F7 \wedge 0.5)$$

and, finally, range is determined in column H by

= (G7/E7)*$F$4

Plot the cruising speed versus density altitude for each power setting, as shown in Fig. 9-15. Fit a linear trendline to these data, and label for the appropriate power settings. Range can also be plotted against altitude, as shown in Fig. 9-16. While there may be a slight curvature to these data points, there is usually so much scatter in the data that a linear fit works best here, also. The resulting graphs should be quite useful, since cruise accounts for the major portion of most flights.

Glide Performance

Glide performance involves measuring the rate of sink, which, as explained in Chap. 8, is similar to measuring rate of climb, since rate of sink is merely negative rate of climb. Timing a descent through an altitude increment is done just like in rate of climb, and the corrections are then quite similar to those for rate of climb. Figure 9-17 shows the spreadsheet for this process.

Enter takeoff weight in cell D4, and fuel burn in E4. Average weight during the tests is calculated in F4. Also enter the calibration factors in G4 and H4. The primary data taken will be the altitude increment through which you descend (Del Hm), which you enter in column C, and the time measured during the descent, to be entered in column D. Also record in column H the indicated airspeed (in knots) held during the run, the pressure altitude in column A, and the OAT (in degrees Fahrenheit) in column B. The example shown was actually measured in miles per hour, and the Vind values corrected to knots; hence the odd airspeed values.

The program then corrects the measured altitude increment (Del Hm) for nonstandard temperature to actual Del Ha in column E, by the formula:

= ((B8 + 460)*C8)/(519*(1 − 0.00000689*A8))

The pressure altitude and temperature from columns A and B are used to calculate density ratio in column F by:

= ((1 − 0.00000689*A8) ^ 5.256)/((B8 + 460)/519)

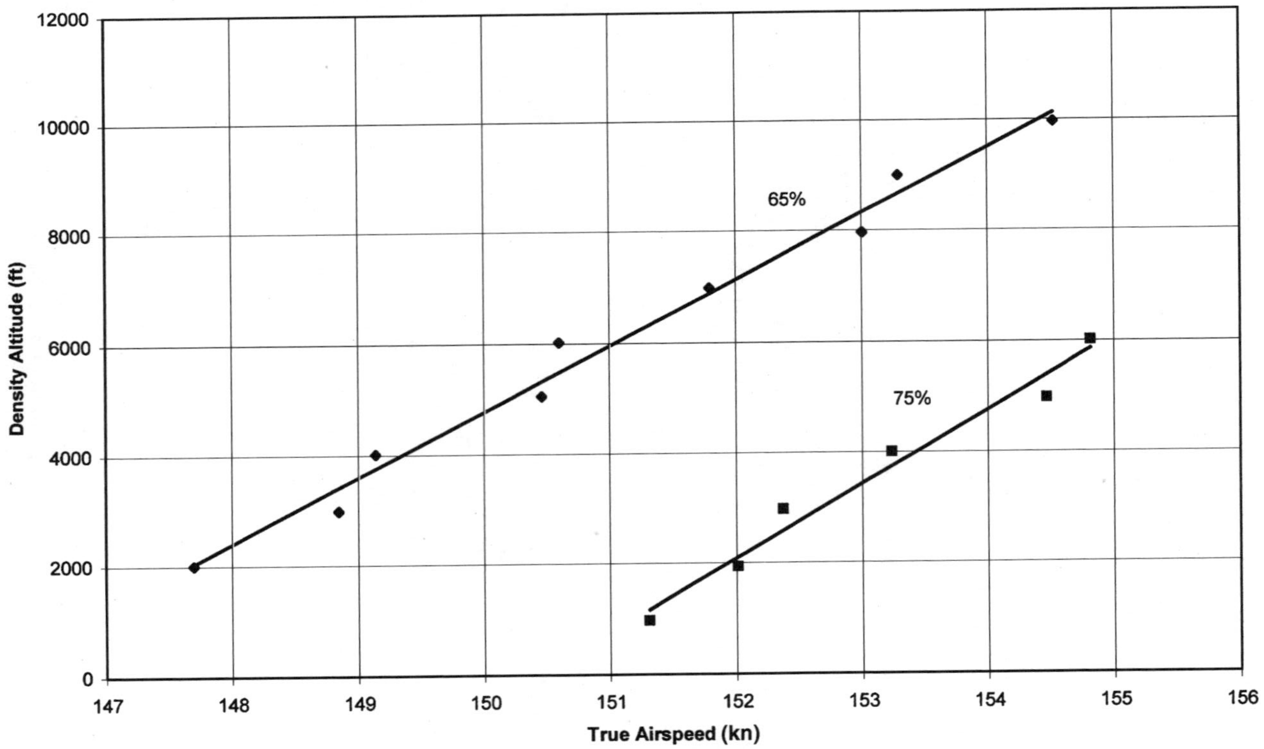

Fig. 9-15. Plots of cruising speed versus altitude at two power settings.

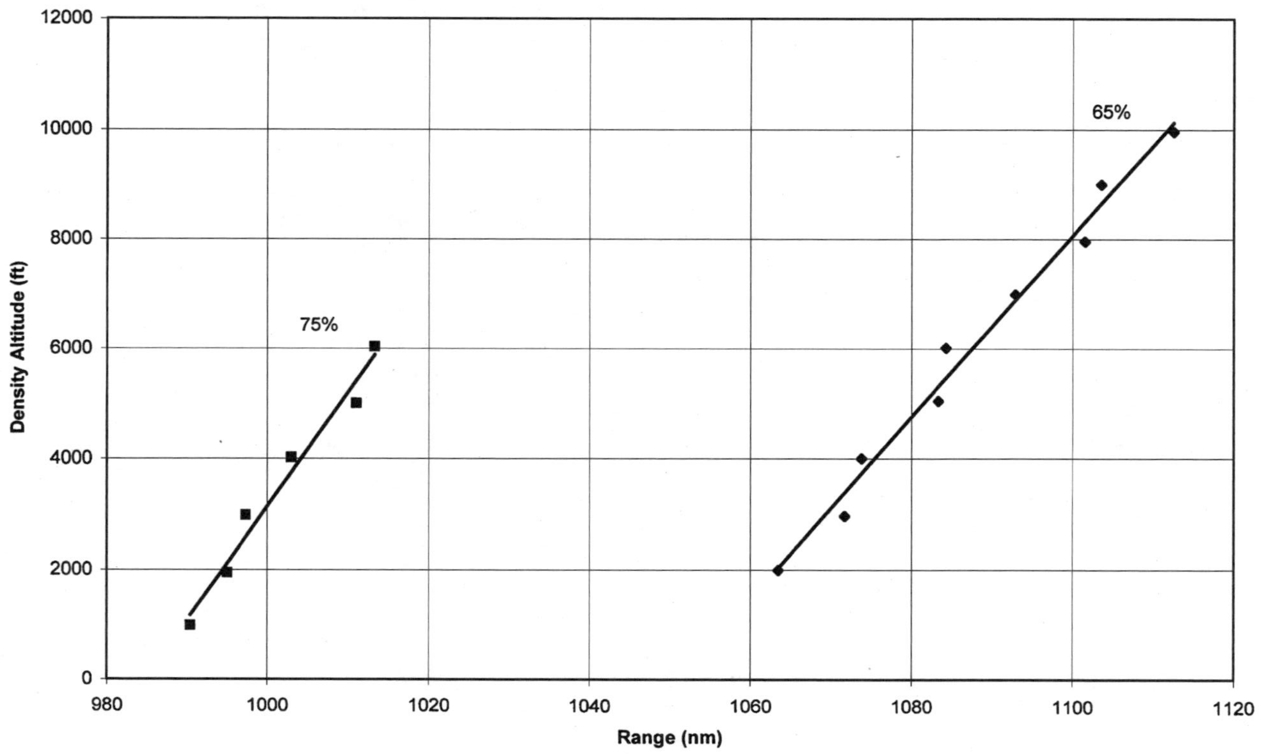

Fig. 9-16. Plots of range versus altitude for two power settings.

	A	B	C	D	E	F	G	H	I	J	K
1	Glide Performance										
2											
3				W to	Fuel Burn	W av	Vcal factors				
4				1785	5	1770	1.007	-2.05			
5											
6	P Alt (ft)	OAT (F)	Del Hm(ft)	Time(sec)	Del Ha(ft)	Dens Rat	D Alt (ft)	Vind (kn)	Vtrue(ft/s)	R/S (ft/s)	L/D
7											
8	5000	50	400	35.05	407.0878	0.846396	5578.023	60.9	108.8883	11.61449	9.375212
9	5000	50	400	30.84	407.0878	0.846396	5578.023	69.5	124.7967	13.19999	9.454303
10	5000	50	400	27.43	407.0878	0.846396	5578.023	74	133.1209	14.84097	8.969827
11	5000	50	400	26.18	407.0878	0.846396	5578.023	78.3	141.0751	15.54957	9.072607
12	5000	50	400	23.14	407.0878	0.846396	5578.023	82.6	149.0294	17.59238	8.471244
13	5000	50	400	22.45	407.0878	0.846396	5578.023	87	157.1686	18.13308	8.667504
14	5000	50	400	19.33	407.0878	0.846396	5578.023	91.3	165.1228	21.05989	7.840628
15	5000	50	400	18.03	407.0878	0.846396	5578.023	95.7	173.262	22.57836	7.67381
16	5000	50	400	14.08	407.0878	0.846396	5578.023	100	181.2162	28.91248	6.267751
17											
18										Max L/D=	9.45
19			Glide Dist (nm)		D Alt (ft)		W (lb)	Vglide(kn)		Vglide=	70
20			0		0		1800	70.2935			
21			3.110599		2000		2000	74.09586			
22			6.221198		4000		2200	77.71239			
23			9.331797		6000		2400	81.16794			
24			12.4424		8000						
25			15.553		10000						

Fig. 9-17. Spreadsheet for calculation of glide performance.

and then density altitude in column G by:

= (1 − F8 ^ 0.235)/0.00000689

The indicated airspeed is corrected through calibrated values (by the Vcal factors) and then directly to true airspeed in column I. This is also calculated here in feet per second, the units needed to determine *L/D* ratio. The formula is:

= (H8*$G$4 + $H$4)*1.69/SQRT(F8)

Rate of sink is then calculated in column J by dividing the values in column E by the respective values in column D, and, finally, *L/D* determined by the ratio of V to R/S (column I divided by column J).

Plot the *L/D* values against Vind, as shown in Fig. 9-18, and fit a polynomial trendline to the data. There will probably be a lot of scatter in the data, but the automatic curve fit of the Chart Wizard should give you a pretty good curve. Note the peak of the curve, which is your max *L/D*, and the airspeed at which this occurs, which is your best glide speed. If done at a low gross weight (as in the example) the peak will occur at a low airspeed. Enter these values on the spreadsheet in cells K18 and K19, respectively.

You can now determine the glide distance from any altitude when flying at the max *L/D*. List the altitudes desired (up to 10,000 ft shown), and the glide distance for each altitude is calculated in column C by:

= $K$18*E20/6076

This formula is for nautical miles. If you want statute miles, simply enter 5280 instead of 6076 in this formula. Plot column C against column E, and fit a linear trendline, as shown in Fig. 9-19. This graph gives still-air glide distance for any altitude. Of course, the density altitude of the terrain must be considered, and subtracted off this distance. For example, if you started a glide at 9000 ft, where the distance shown is about 14 mi, but the density altitude of the ground level is 2000 ft, you would have to subtract about 3 mi, to yield a net range of 11 mi.

As explained in Chap. 8, the *L/D* ratio, and, hence, glide distance, is not affected by weight or altitude. The *velocity* for max *L/D*,

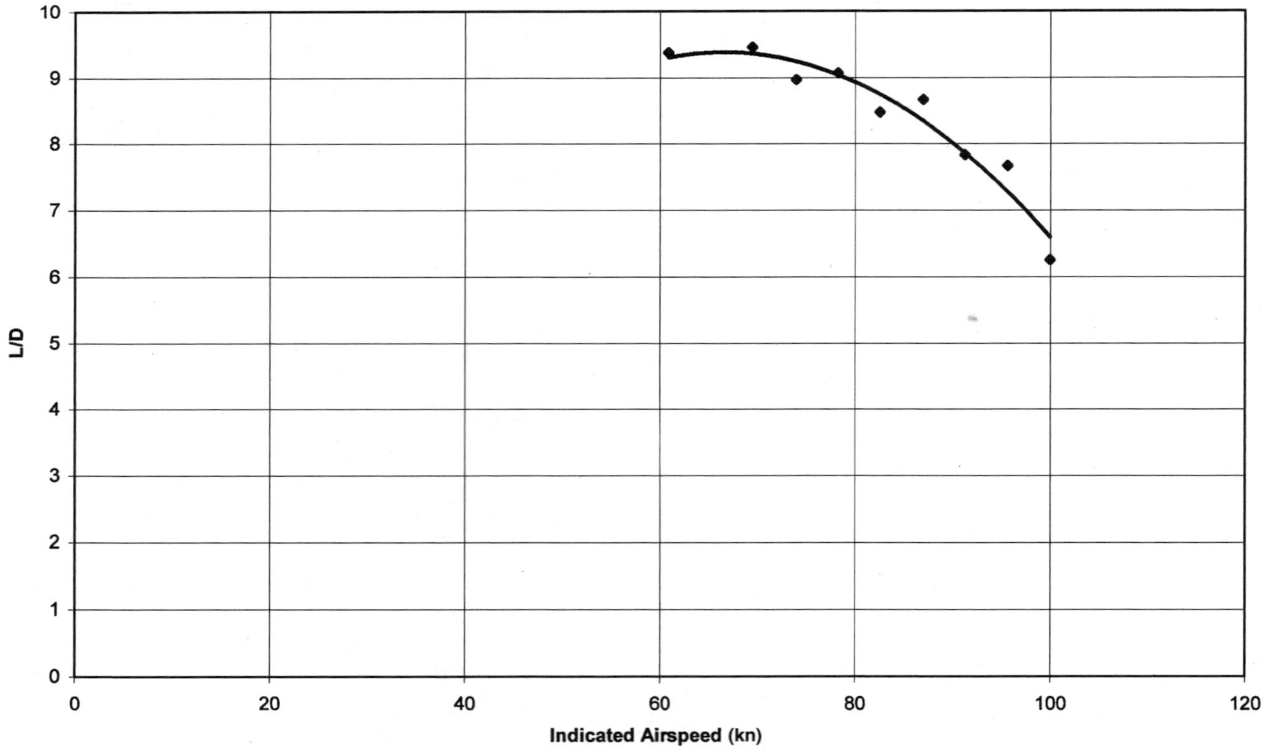

Fig. 9-18. Plot of lift to drag ratio versus airspeed.

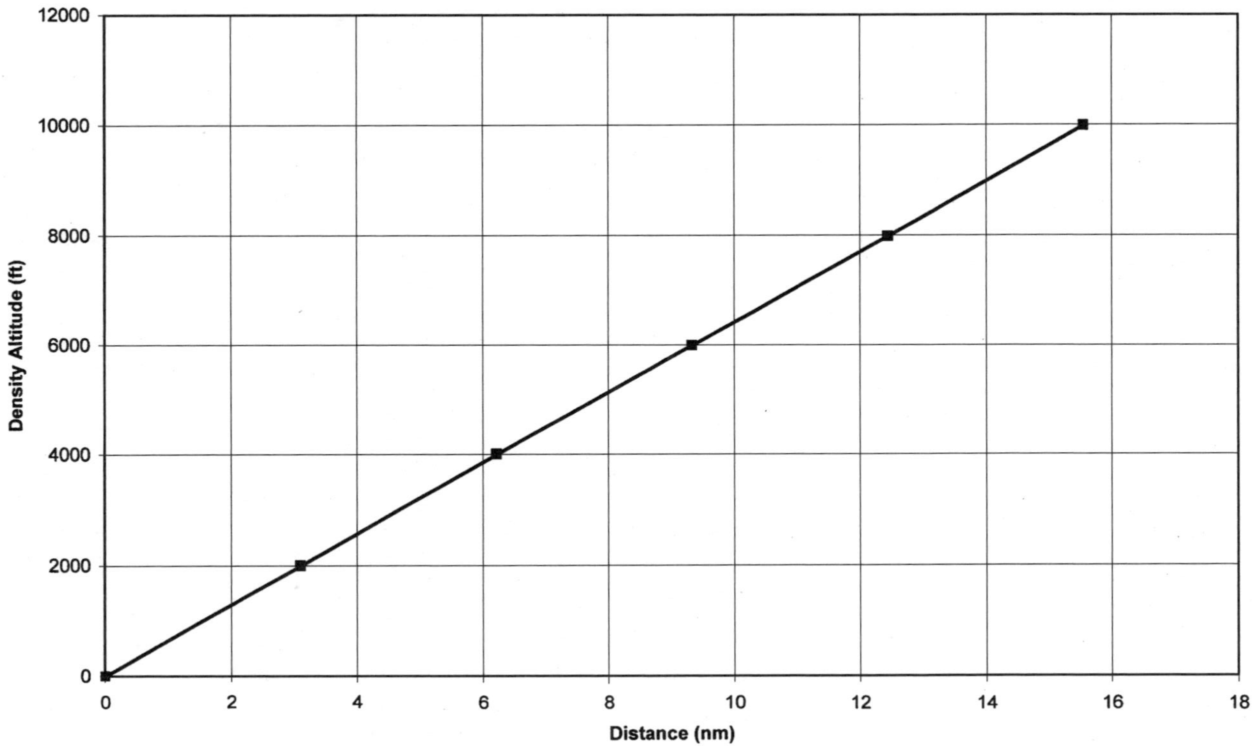

Fig. 9-19. Plot of maximum glide distance versus altitude.

however, is dependent on weight. The best glide speed for the test weight here (1770 lb) is 70 kn indicated, as shown in K19. To determine best glide speed for other gross weights, enter the desired weights in column G, beginning in row 20. Column H, beginning in row 20, then calculates best glide speed for the respective weights by:

= $K$19*SQRT(G20/$D$4)

You could record these values on Fig. 9-19, or list them elsewhere, at your discretion. Round them off, however, to the nearest knot. For the example, 2400 lb is the maximum gross weight for the airplane tested, and the resulting best glide speed would be 81 kn.

Altitude also affects the best glide speed, but that is the *true* airspeed. If calibrated airspeed is used, the best glide speed is the same for all altitudes (for a given weight). Since this is pretty much proportional to indicated airspeed, and that is the actual value you look at, indicated airspeed is used for best glide speed calculations.

Landing Distance

As pointed out in Chap. 8, due to the large variation in pilot technique and braking action, landing roll distance can really only be determined by spotting the touchdown and stopping points, and measuring the distance between them. While there is no calculation involved in this process, there is in the reduction of the test conditions to standard, usually, standard weight and standard sea level. Wind effects may also need to be eliminated. The spreadsheet in Fig. 9-20 provides a program for these corrections. Data required includes the standard weight (usually max gross) in cell A4, the takeoff weight during the test in B4, the fuel burn in gallons during the test in C4, the pressure altitude in E4, the outside air temperature in F4, the headwind (assuming you land into the wind) in G4, and the indicated stall speed (for the test gross weight). Also, enter the airspeed calibration factors from your calibration curve in A7 and B7.

The program will then calculate the average weight during the test in D4, from the formula:

= (B4 + (B4 − 6*C4))/2

	A	B	C	D	E	F	G	H
1				Landing Distance				
2								
3	Wstd (lb)	Wto	Fuel Burn	Wav (lb)	Press Alt	OAT (F)	Hdwind	Vsi (kn)
4	2400	2165	5	2150	1500	68	7	47
5								
6	Vcal factors		L/D			Dens Rat	Vst (kn)	Vldg (kn)
7	1.007	-2.05	6.9			0.930721	46.93395	53.97404
8								
9	d(land)=	625	d(0wind)=	808.137	d(stdsl)=	752.15	d(stdwt)=	937.2383
10								
11		D Alt	Dens Rat	dstdw	d2000	dairstdw	dtotstdw	
12		0	1	937.2383	650.8599	574.7358	1511.974	
13		1000	0.971003	965.2266	670.2963	591.8989	1557.126	
14		2000	0.942654	994.2544	690.4544	609.6994	1603.954	
15		3000	0.914943	1024.368	711.3666	628.1657	1652.534	
16		4000	0.887859	1055.616	733.0665	647.3276	1702.943	
17		5000	0.861393	1088.049	755.5899	667.2166	1755.266	
18		6000	0.835535	1121.723	778.974	687.8658	1809.588	
19								
20								
21								
22								
23								

Fig. 9-20. Spreadsheet for landing distance calculation.

the density ratio for the test in F7 by:

$$= ((1 - 0.00000689*E4) \wedge 5.256*519/(F4 + 460)$$

and the stall speed in true airspeed in G7 by:

$$= (H4*A7 + B7)/SQRT(F7).$$

Landing speed is then determined in H7 as 1.15 times the stall speed in G7. If you want to calculate the landing distance for full-stall landings simply replace the 1.15 in this formula with 1.0.

Now enter the measured landing roll in feet B9 (625 ft in the example). The program will proceed to correct this value for wind in D9 by:

$$= B9*((H7/(H7 - G4)) \wedge 1.85)$$

then correct that value to standard sea level in F9 multiplying D9 by F7, and finally correct that to standard weight in H9 with the formula:

$$= F9*(A4/D4) \wedge 2$$

This value (937 ft in the example) is your stopping distance for zero wind at standard sea level and standard weight. The rest of the program is available to calculate landing distance at higher altitudes or reduced gross weight. For this procedure, list the desired altitudes in column B beginning in row 12. Altitudes to 6000 ft are listed in the example. The density ratio for each altitude is calculated in column C by:

= (1 − 0.00000689*B12) ^ 4.256

The sea level value is copied from H9 into D12, and this is increased for higher altitudes by the formula:

= $D$12/C13

and copied on down the column. Column E provides the same computation for a reduced gross weight, 2000 lb, by first correcting the sea level value with the formula:

= D12*((2000/$A$4) ^ 2).

For any other desired weight, just substitute that value for the 2000. This distance is increased in the succeeding rows of that column by:

= $E$12/C13.

All of the above discussion pertains to the ground roll portion of the landing. If you want to calculate the air distance to add to this, i.e., the distance to descend from 50 ft to the touchdown point, you must first determine the *L/D* for the glide as described previously. For landing performance, though, this test would have to be done in landing configuration: landing gear down and flaps down. Once this test is accomplished, and the resulting curve plotted, enter the curve at the average approach speed, and read the *L/D.* This speed would be the average of your speed at 50 ft (1.3 Vs) and your touchdown speed. Using 1.15 Vs for touchdown, as is done in the program, the average value would be 1.225 Vs. This value would also have to be adjusted for the appropriate gross weight, as described in the glide distance section.

Enter the resulting *L/D* value in cell C7. The program will then calculate the distance to descend from 50 ft in column F by the formula:

= $C$7*((50 + 0.01624*(($H$4*$A$7 + $B$7)) ^2)/C12).

The distances in column F are then added to the ground roll distances in column D to yield total landing distances over a 50-ft obstacle for the various altitudes in column G. A plot of both the ground roll distances and the 50-ft-obstacle distances is shown in Fig. 9-21. The chart shows only the values for standard, or maximum gross, weight. Fit a polynomial trendline to each of the curves for a usable chart.

Conclusion

These procedures will enable you to convert the raw data you take in flight into useful charts with a minimum of calculating effort. All of the basic performance items in the preceding chapters are covered. If you are proficient in working with spreadsheets, you may want to create additional ones, or modify these for your own personal use. To avoid confusion, the numbers resulting from various calculations have been reproduced exactly as they appear on the spreadsheet, which, in many cases is to four decimal places. For practical purposes, most of these can be rounded off to whole numbers, or, at least, to one or two decimal places.

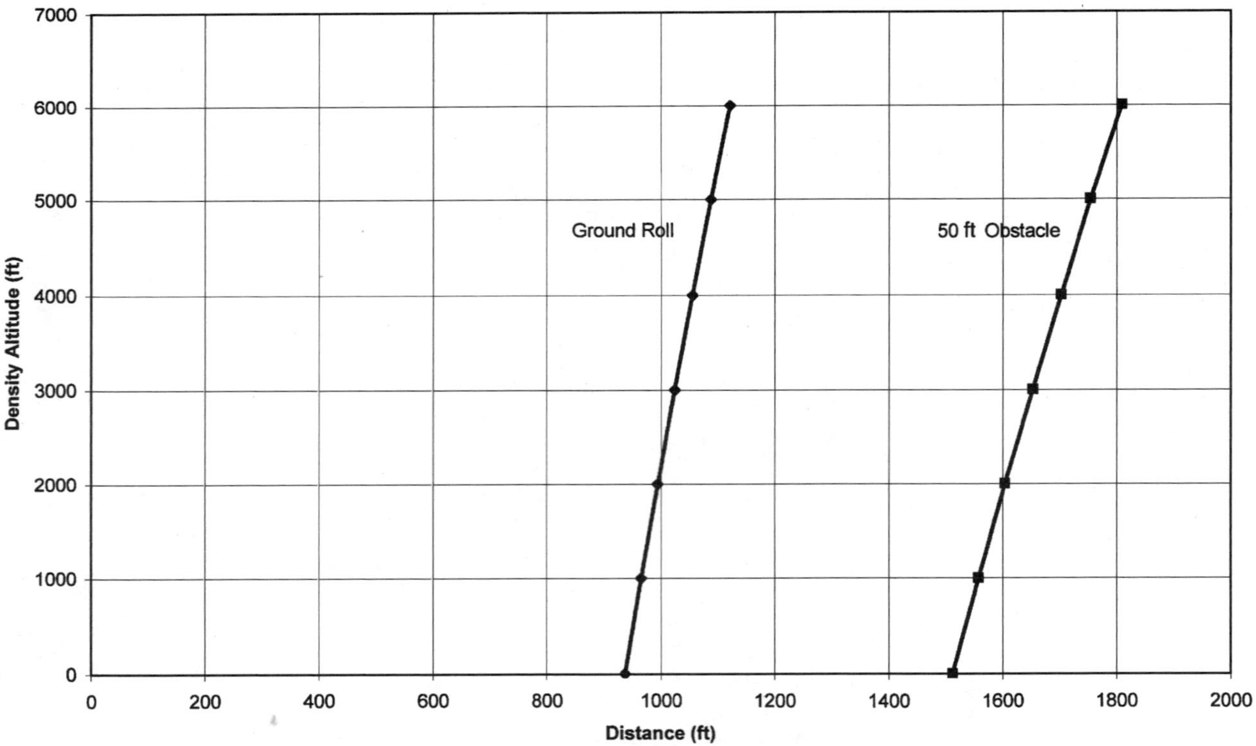

Fig. 9-21. Plots of landing ground roll distance and landing distance over 50-ft obstacle versus altitude.

CHAPTER 10

Evaluating Modification Effects

Many basic aircraft designs have not changed much in 30 years or more. However, in that time, significant progress has been made in improved aerodynamics, particularly in small details of airplane design. Consequently, numerous aerodynamic modifications have been developed as after-market devices to install on existing airplanes. These modifications include such things as control-surface gap seals, hinge fairings, landing gear fairings, redesigned wingtips, and other such devices. They have had such appeal that a number of companies have been formed that are devoted solely to the production and marketing of these modifications. One company has delivered almost 30,000 kits since their inception in the early 1980s, and currently sell about 2500 each year.

Most of these kits are for modifications that are designed to increase aerodynamic performance. This improvement might be seen in higher cruise speed, greater rate of climb, lower stall speed, or in better handling. Some account for only a minor improvement when considered individually, but, when added to others, make a significant difference. Others can make a measurable change by themselves. Owners who install such devices, are, understandably, concerned about the amount of improvement that they provide. The question then arises as to how to best evaluate the effect that these modifications have on an existing airplane's performance.

The simple answer to this question would be to run through all the flight test experiments outlined in the previous chapters before the modification, and then repeat them afterwards. Obviously, this procedure would give you the best overall picture. Certain modifications, however, may not affect all aspects of performance, or the owner may not be concerned about all of them. Someone attempting to increase cruise speed, for example, may not care that the modification for this end will also shorten takeoff by a few feet, or increase range by a few miles. This chapter is included, therefore, to provide some quick evaluations of the effects of modifications on specific performance items.

(a)

Fig. 10-1. Piper Arrow wing lower surface, (*a*) in original configuration.

Speed

By far, the majority of after-market aerodynamic modifications are installed to increase the cruising speed of the airplane. This goal is accomplished by devices that reduce the drag of the airplane, such as gap seals and hinge fairings, as shown in Fig. 10-1 on a Piper Arrow. Numerous small details of the airplane produce unnecessary drag. Smoothing them over, or sealing leakage of air can reduce drag to some degree. With the same power, and, hence, the same thrust, the aircraft will go a little faster.

The measure of just how much faster can be done much as described in Chap. 7. You must, of course, measure the cruising speed at the desired power settings with the aircraft in the original configuration. After modification, it is only necessary to measure the new performance at a few test points to see just how much effect this change has had. Even if you haven't run extensive flight tests at various altitudes, and constructed a speed-altitude curve for a specific power setting, you could just choose a few points for before and after testing to quickly evaluate the results. In fact, this goal could be accomplished without any previous flight testing, even airspeed calibration. The Global Positioning System (GPS) can be used to get accurate airspeed measurements.

The key to this measurement is to make sure that you measure both the before and after performance at the same density altitude and power setting. You don't even have to know the exact power setting, as long as it is the same, and at the same density altitude, for the before and after tests, such as a given rpm, or combination of rpm and manifold pressure in the case of a constant speed prop. Flying at the same density altitude will assure that the aerodynamic effects will be similar, and also that the power for a given setting will be the same. The same power means the same thrust, and any change in airspeed will, hence, be due to change in drag.

(b)

Fig. 10-1. (*Continued*) Piper Arrow wing lower surface (*b*) with gap seals and hinge fairings installed.

To illustrate this procedure, let's consider an example. Prior to the modifications, we fly the airplane to 5000-ft pressure altitude, and note that the OAT is 50°F. From Fig. 10-2 (similar to Fig. 1-3), we can see that this combination of atmospheric conditions gives 5500-ft density altitude. Now, we set the power to a desired setting. Assume that the airplane is a Piper Arrow, and we could, thus, consult the power chart in Table 7-1. We choose 65 percent power, which requires 21.6 in Hg at 2400 rpm. Also, we need to lean the mixture to a specific point, say peak EGT.

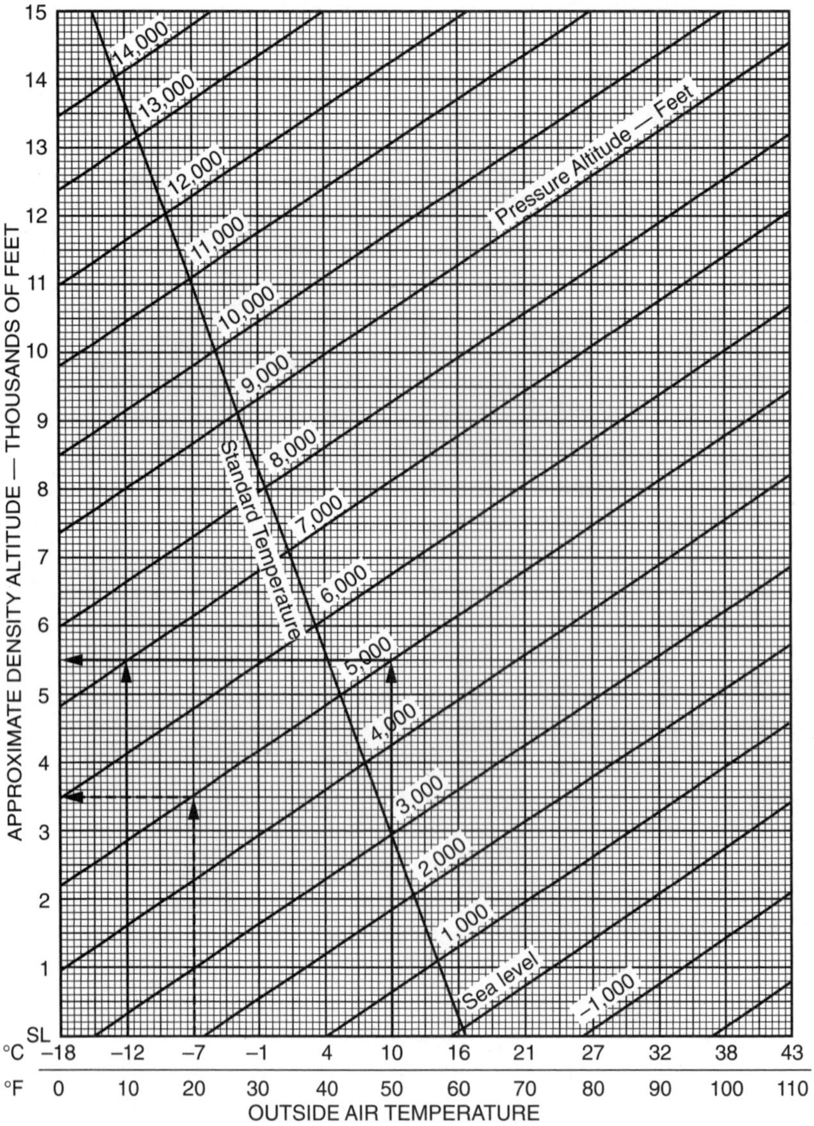

Fig. 10-2. Density altitude chart showing original tests at 5000 ft and 50°F, and after mod tests at 7000 ft and 10°F, both yielding the same density altitude of 5500 ft.

With that power setting, you can proceed to measure airspeed by the GPS Method outlined in Chap. 3. Fly the airplane into the wind, and, when steady, read the groundspeed. Then head downwind, and take another reading. Since you are looking for true airspeed here, simply add these two values and divide by 2. You could try another power setting or so, if desired. That completes the tests in standard configuration.

Now, the airplane is put into the shop, and the desired modifications performed. When completed, we run the tests for the modified configuration. We fly to 5000-ft pressure altitude again, but note that on this day, the OAT is only 20°. Consulting our chart (Fig. 10-2), we find that this combination gives only 3500 ft density altitude. We must continue to climb, and check the temperature against altitude to find density altitude. At 7000 ft, we note that the OAT is 10°, giving just about 5500-ft density altitude, the same as in the original tests. Now we set the same power settings and mixture as before, and proceed to repeat the tests, measuring groundspeed in both directions on the GPS, and averaging these readings to yield true airspeed. The difference from our original reading is the actual speed increase, reflecting the drag decrease.

Note that we would not need to set exactly 65 percent power. We could simply have chosen, for example, 20 in Hg and 2400 rpm. If you were testing an aircraft with a fixed pitch prop, you could maybe select 2400 rpm, and possibly repeat the test at 2300 rpm. The trick is to make sure that you record these settings, and set them again for the after-modification tests. We have also not relied on the airplane's airspeed system to give the speed readings. You could also note the indicated airspeed in each test, and then correct these to true airspeed for comparison with the GPS results.

Climb Performance

If rate of climb is of importance to you, this performance could also be measured without resorting to the extensive procedure required to get the complete climb schedule up to absolute altitude, as described in Chap. 6. If you are flying a production aircraft, you should know the recommended speed for best rate-of-climb V_y. To determine the change in rate of climb after a modification, you could use this speed for measuring performance before and after the modification. Run the test as described in Chap. 6, but only at the one speed. You could also use just one altitude, or, maybe one or two more. Make sure you know the *density* altitude, though, and then repeat the test after the modification at this same density altitude.

If flying a homebuilt, or other aircraft for which you don't have very reliable performance information, you may have to do a more extensive test using several velocities, in order to find V_y. Again, one altitude would be sufficient. Once V_y is determined in the before-modification tests, you could use that speed, and just fly at it for the after-modification check. It is not really necessary to perform the tests at the best climb speed, but that would give you an indication of the maximum performance. If you don't want to take the time to find it, you could choose an arbitrary indicated airspeed, and fly both before and after tests at that value.

As noted in Chap. 6, very still air is hard to find, but it is required for reliable vertical speed tests. It is usually necessary to repeat each data point several times in measuring rate of climb. Keep repeating runs until you get consistent times for climb at the same velocity and altitude.

There should be negligible change in best climb speed for minor aerodynamic changes, and also small change in the rate of climb. This would not be the case if you were making a major change, such as the installation of a more powerful engine. In that case, the entire climb performance schedule may change significantly. It is recommended, then, that the complete set of tests be run according to procedures in Chap. 6, both before and after the modification. In fact, it would be a good idea to do this with all performance tests, since, in this case, you really have a whole new airplane. Of course, if you want to determine the ceiling, you would also have to do the complete set of climb tests.

Stall Speed

Recently, there has been a trend to install vortex generators on the wings in order to lower the stall speed, as shown in Fig. 10-3. These devices create artificial turbulence, which energizes the air in the boundary layer over the wing, and delays the separation of the airflow at high angles of attack until an even higher angle is attained. This whole process results in lower stall speed. Lower stall speed means shorter takeoff and landing distances, lower minimum control speed for twins, and other performance advantages. Some modifications

Fig. 10-3. Micro vortex generators installed on a Piper Cherokee wing. (*Courtesy Micro Aerodynamics, Inc.*)

aimed at drag reduction, also, may change stall speed slightly, in some cases making it higher.

If, for any of these reasons you want to check the effects of modifications on stall speed, it is relatively simple. The stall speed is not dependent on altitude when measured in calibrated airspeed, which is usually very close to indicated airspeed. Hence, flying before and after stall tests should give a pretty realistic comparison just by checking IAS for stall in both conditions, and at any altitude. Weight *does* affect stall speed, however. Fortunately, there is a specific relationship here, though. As pointed out in Chap. 4, the stall speed varies as the ratio of the square root of the weights. If, for some reason, you can't fly at nearly the same weight for the before and after tests, you can correct for the difference with the equation,

$$V_{S_2} = V_{S_1}\sqrt{\frac{W_2}{W_1}}$$

Perhaps, you want to know the effect of lowered stall speed on takeoff and landing performance. If you have already done these tests, and have pretty good data, and, if you have made no other major changes to the airplane, the new distances could be computed rather simply. Remember that the takeoff speed is defined as a percentage above stall speed (usually 20 percent). Hence, if there is no significant change in drag or other characteristic of the aircraft, the new takeoff distance could be computed by entering the chart shown in Fig. 5-10 at the new takeoff speed (1.2 times V_S), and reading the corresponding distance. You would have to run through the corrections to standard conditions, though, since your original tests were probably performed at other than standard. This process would be especially easy if you used the computer program in Chap. 9.

The air distance (glide from 50 ft) portion of landing distance could also be adjusted by just changing the value of stall speed in the equation in Chap. 4. The ground distance, or braking distance, however, would have to be measured again by the method described. There would be some change in it, though, because the touchdown speed is also a function of stall speed (normally, 1.15 V_S).

One last comment is in order here. Lowering stall speed with vortex generators increases the drag on the wings. The amount of this increase may not be significant in some cases, but, in others, it could be. If so, it might be prudent to check the other performance of the airplane, when this modification is done. Items such as rate of climb, cruise speed, range, and glide distance, among others could be affected.

APPENDIX A

Symbols, Conversion Factors, Data Sheet

List of Symbols

bhp	brake horsepower
CAS	calibrated airspeed
D	drag
d	distance
d_A	air distance (takeoff or landing)
d_g	ground distance (takeoff or landing); also glide distance
d_0	distance with zero wind
d_{SSL}	distance at standard sea level
E	endurance
EGT	exhaust gas temperature
FC	fuel consumption
GR	glide ratio
GS	groundspeed
hp	horsepower
h	altitude
IAS	indicated airspeed
K	general constant

L	lift
L/D	lift-to-drag ratio
MAP	manifold air pressure
P	pressure; also, power
R	range
RC (or R/C)	rate of climb
rpm	revolutions per minute
RS (or R/S)	rate of sink
S	wing area
T	temperature; also, thrust
t	time
TAS	true airspeed
thp	thrust horsepower
V	velocity
V_c	cruise velocity
V_{cal}	calibrated airspeed (same as CAS)
V_{G}	groundspeed; also, glide speed
V_i	indicated airspeed (same as IAS)
V_L	landing velocity
V_{to}	takeoff velocity
V_S	stall velocity
V_t	true airspeed (same as TAS)
V_x	best angle-of-climb velocity
V_y	best rate-of-climb velocity
W	weight
α	angle of attack
Δ (or Del)	difference
ρ	density
ρ_0	standard sea level density

Conversion Factors

MULTIPLY	BY	TO GET
Miles per hour (mph)	0.8696	Knots
Knots (kn)	1.1508	Miles per hour
Miles per hour (mph)	1.4667	Feet per second
Knots (kn)	1.6879	Feet per second
Miles (U.S.)	0.8696	Nautical miles
Nautical miles (nm)	1.1508	Miles (U.S.)
Miles	5280	Feet
Pounds/cu inch	2.0360	Inches of Hg
Inches of Hg	0.4912	Pounds/cu inch

Temperature:

$$°C = (°F - 32)\left(\frac{5}{9}\right) \qquad °F = \left(\frac{9}{5}\right)(°C) + 32$$

Metric Conversion Factors

MULTIPLY	BY	TO GET
Feet	0.3048	Meters
Gallons	3.7853	Liters
Horsepower	0.7457	Kilowatts
Knots	0.5151	Meters per second
Miles per hour	0.4471	Meters per second
Pounds	4.448	Newtons
Pounds per sq. inch	6895	Pascals

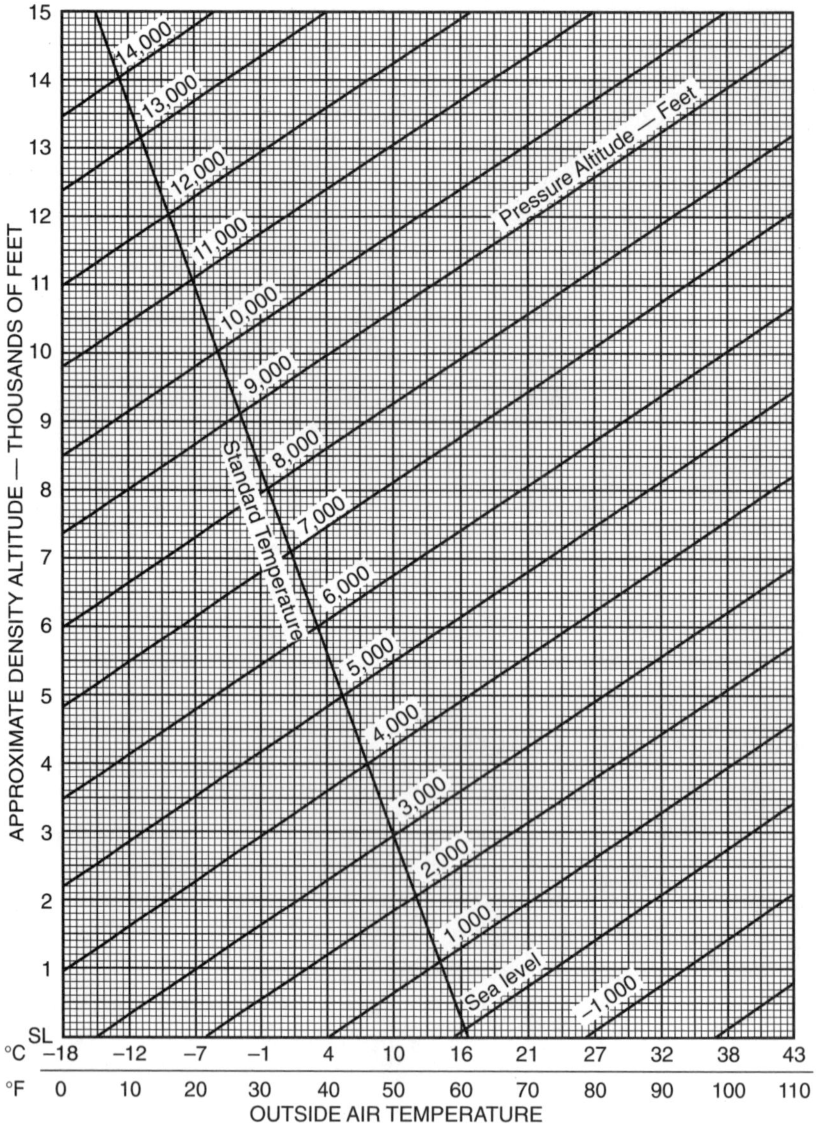

Fig. A-1. Chart for determining density altitude from pressure altitude and temperature.

FLIGHT TEST DATA

TEST: ______________________
DATE: ______________________
LOCATION: ______________________
SL PRESS: ______________________
SURFACE TEMP: ______________________
PILOT: ______________________
OBSERVERS: ______________________
______________________ FUEL______
____

WEIGHT
EMPTY ________
FUEL ________
PASS. ________
BALLAST ________
TOTAL T.O ________
LANDING ________
AVERAGE ________

RUN	IAS	P ALT	OAT	TIME	POWER		

REMARKS:

Fig. A-2. Typical flight test data sheet.

APPENDIX B

Performance Data Examples

This appendix contains actual performance data as presented in *Pilot's Operating Handbooks.* Stall, takeoff, climb, cruise, descent, and landing performance charts are included for the Piper Archer II (Figs. B-1 to B-18). The charts are typical of performance information presented in graph form. The vertical scale in most of these graphs is density altitude. Notice that instead of simply labeling the scale in density altitude, a chart for determining density altitude from pressure altitude and temperature is provided along the left hand side of each graph. The intersections of the standard temperature line and the various pressure altitudes would yield respective values of density altitude.

Performance in tabular form is shown for the Cessna 152 (Fig. B-19 to B-24). These tables also contain performance information in takeoff, climb, cruise, and landing. Interpolation is required for values between figures listed and the notes must be referred to for adjustment to special conditions.

All of this information is provided for illustrative use only. None of it should be used for the actual operation of airplanes involved, since

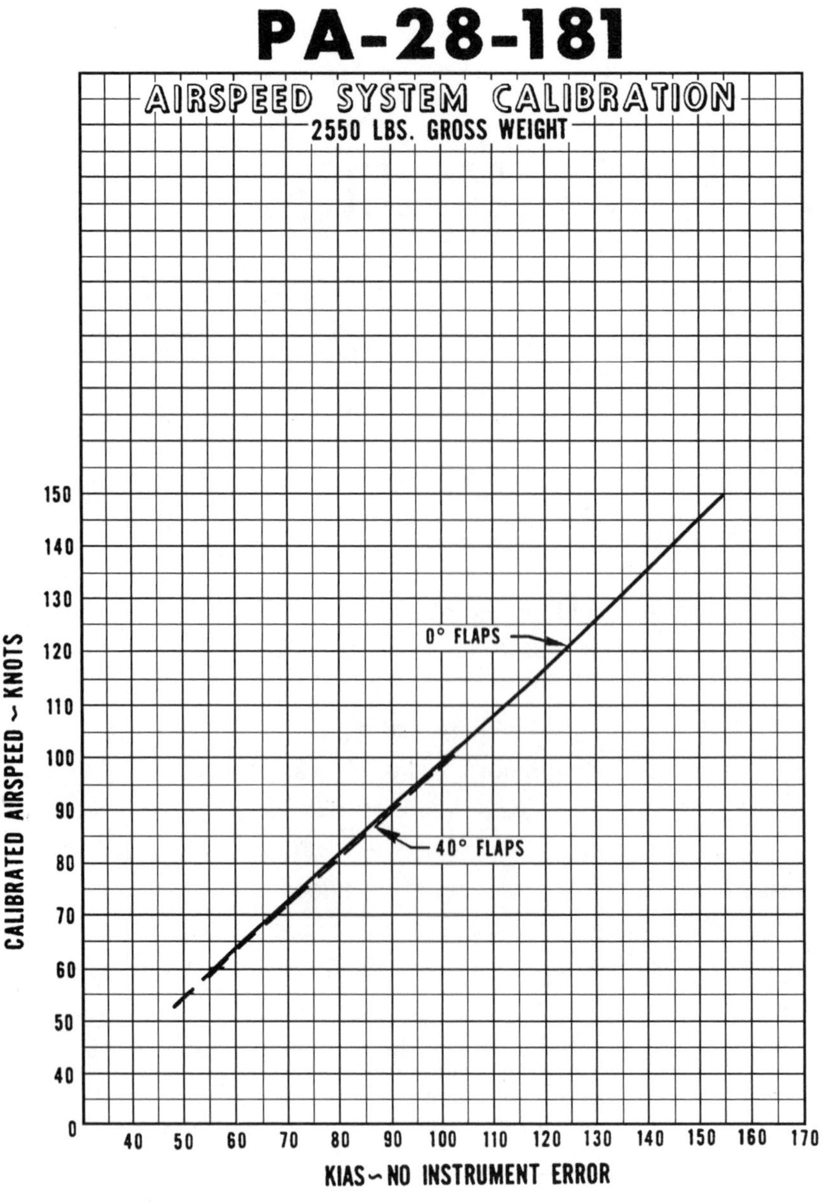
PA-28-181
AIRSPEED SYSTEM CALIBRATION
2550 LBS. GROSS WEIGHT
0° FLAPS
40° FLAPS
CALIBRATED AIRSPEED ~ KNOTS
0
40
50
60
70
80
90
100
110
120
130
140
150
40
50
60
70
80
90
100
110
120
130
140
150
160
170
KIAS ~ NO INSTRUMENT ERROR

Fig. B-1.

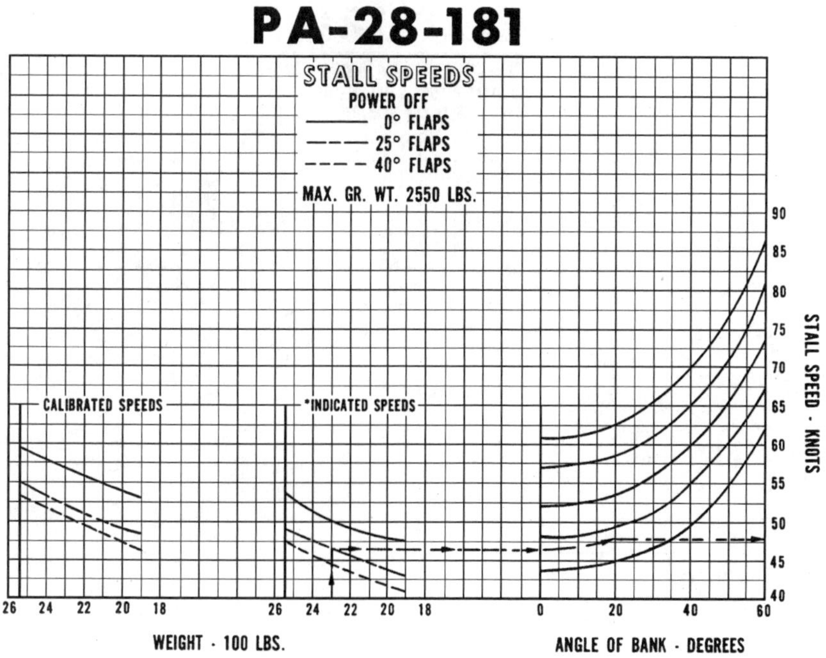

Example:
Gross weight: 2300 lbs.
Angle of bank: 20°
Flap position: 25°
Stall speed: 46 knots (*indicated airspeed)

Fig. B-2.

it is not kept current by appropriate revisions. *Neither the aircraft manufacturers (The New Piper Aircraft, Inc. and Cessna Aircraft Company) nor the author and publisher will be responsible for the accuracy of the data in this appendix.*

This information is provided through the courtesy of The New Piper Aircraft Inc. and Cessna Aircraft Company.

PA-28-181

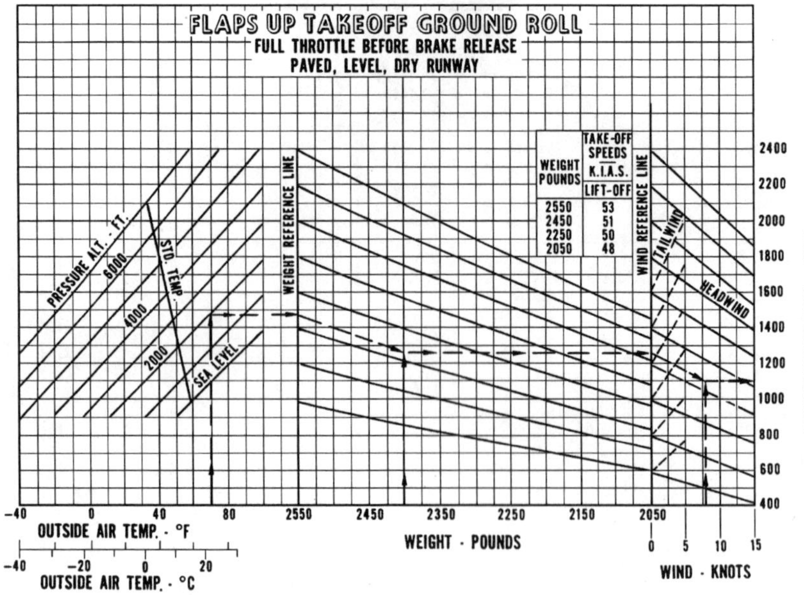

Example:
Departure airport pressure altitude: 2000 ft.
Temperature: 70°F
Gross weight: 2400 lbs.
Wind: 8 knots (headwind)
Takeoff ground roll: 1100 ft.

Fig. B-3.

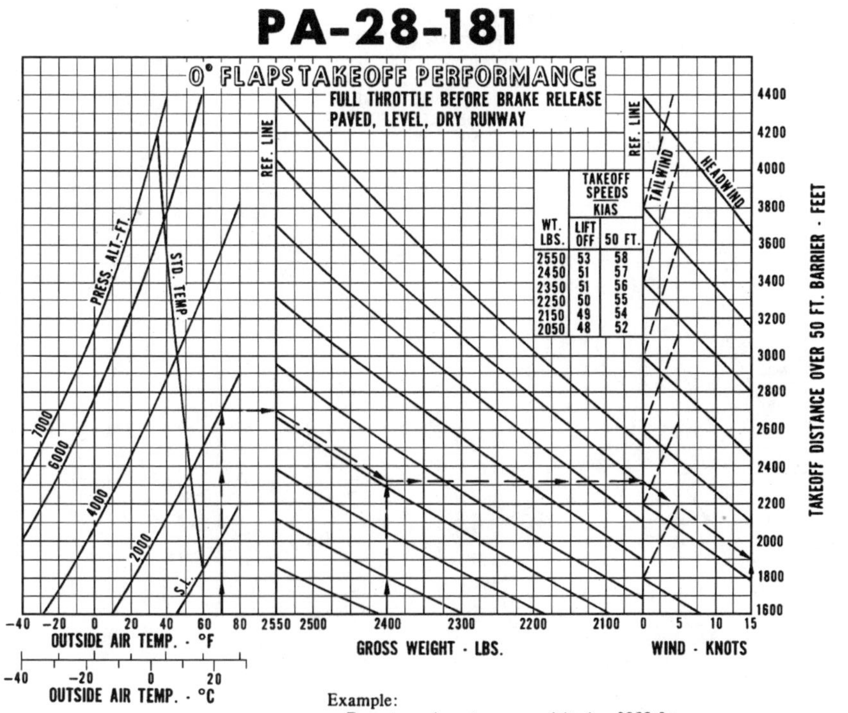

Example:

Departure airport pressure altitude: 2000 ft.
Temperature: 70°F
Wind: 15 KT. (headwind)
Gross weight: 2400 lbs.
Takeoff distance: 1900 ft.

Fig. B-4.

PA-28-181

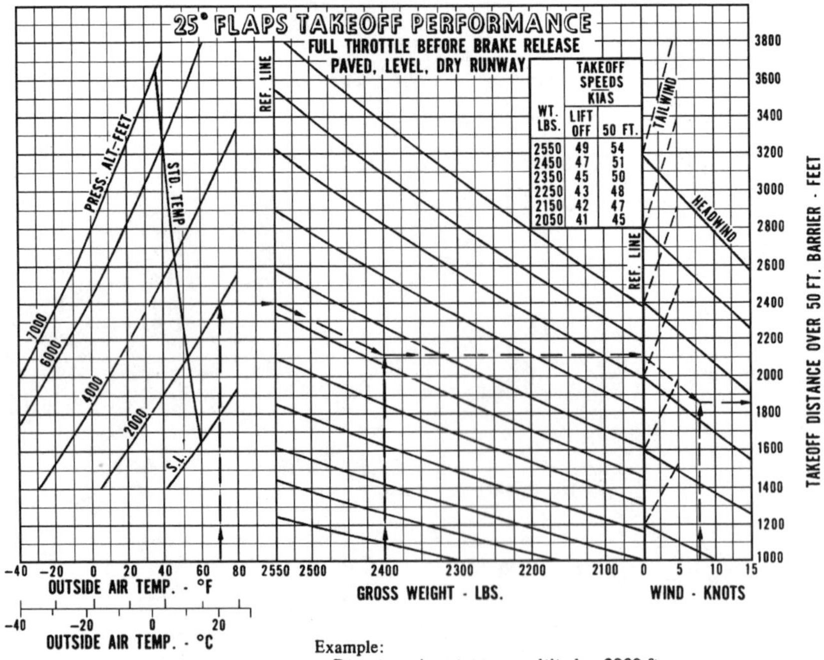

Example:

Departure airport pressure altitude: 2000 ft.
Temperature: 70°F
Gross weight: 2400 lbs.
Wind: 8 knots (headwind)
Takeoff distance: 1860 ft.

Fig. B-5.

PA-28-181

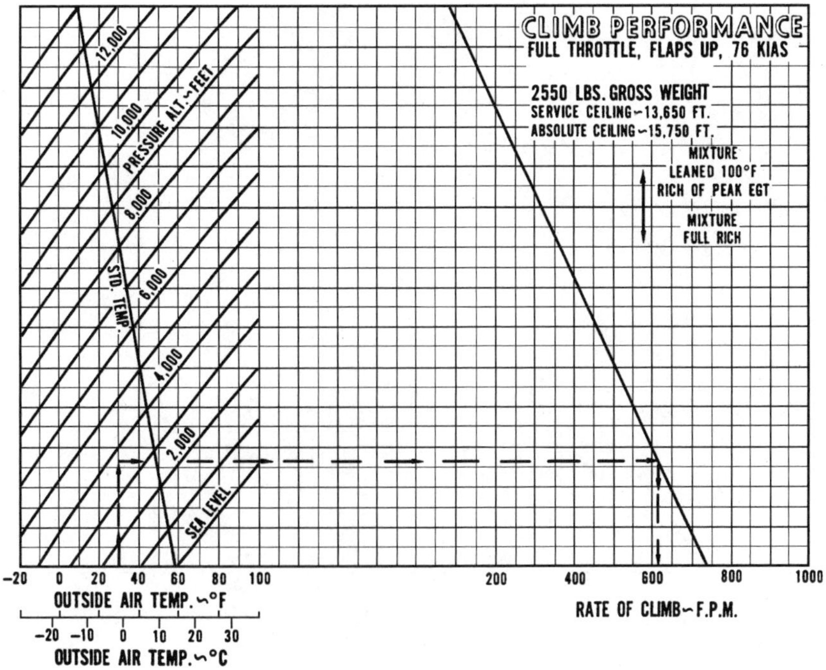

Example:
Climb pressure altitude: 3600 ft.
OAT: 30°F
Rate of climb: 620 F.P.M.

Fig. B-6.

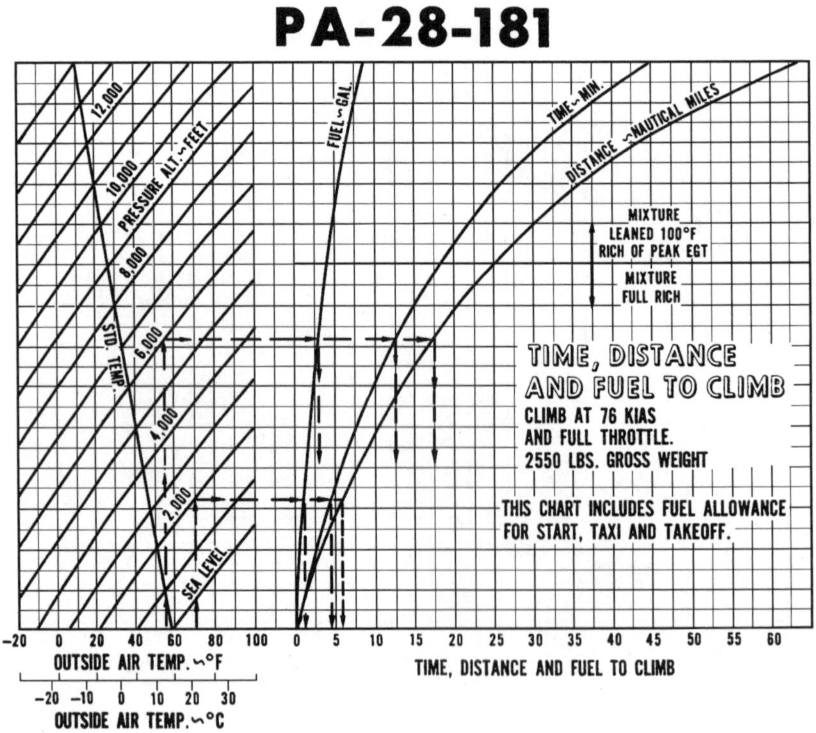

Example:

Departure airport pressure altitude: 2000 ft.
Departure airport temperature: 70° F
Cruise pressure altitude: 6000 ft.
Cruise OAT: 55°F
Time to climb: 12.5 min. minus 4.5 min. = 8 min.
Distance to climb: 17.5 miles minus 6.5 miles = 11 nautical miles
Fuel to climb: 3 gal. minus 1 gal. = 2 gal.

Fig. B-7.

PA-28-181

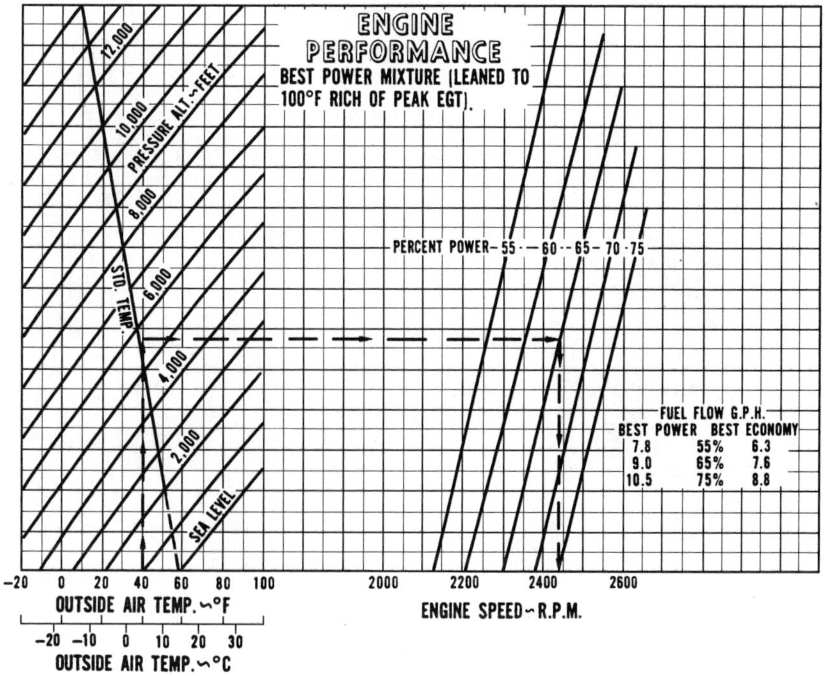

Example:
Cruise pressure altitude: 5500 ft.
Cruise OAT: 40°F
Percent power: 65%
Engine RPM: 2440 RPM

Fig. B-8.

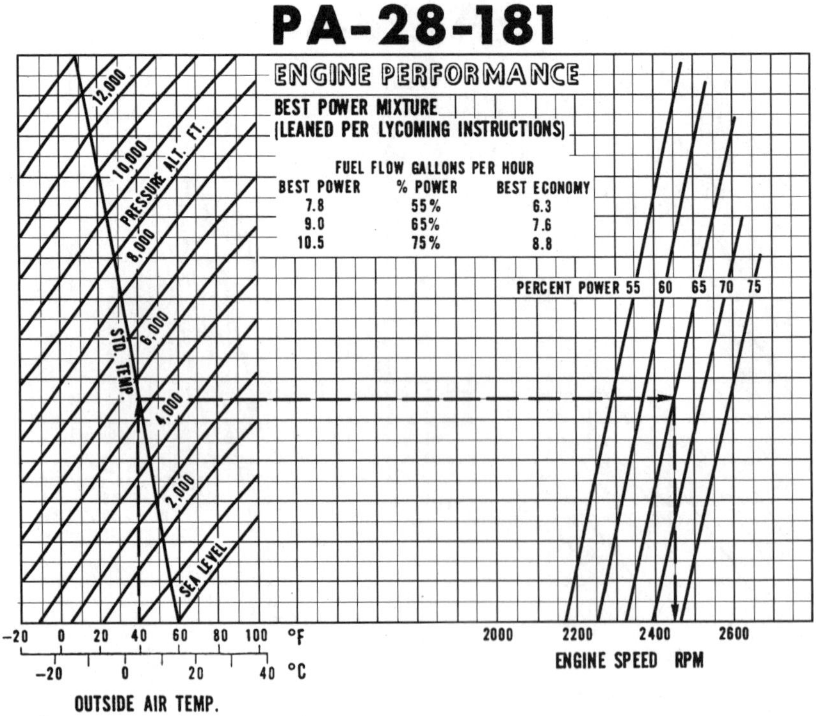

Example:
Cruise pressure altitude: 5500 ft.
Cruise OAT: 40°F
Percent power: 65%
Engine RPM: 2450 RPM

Fig. B-9.

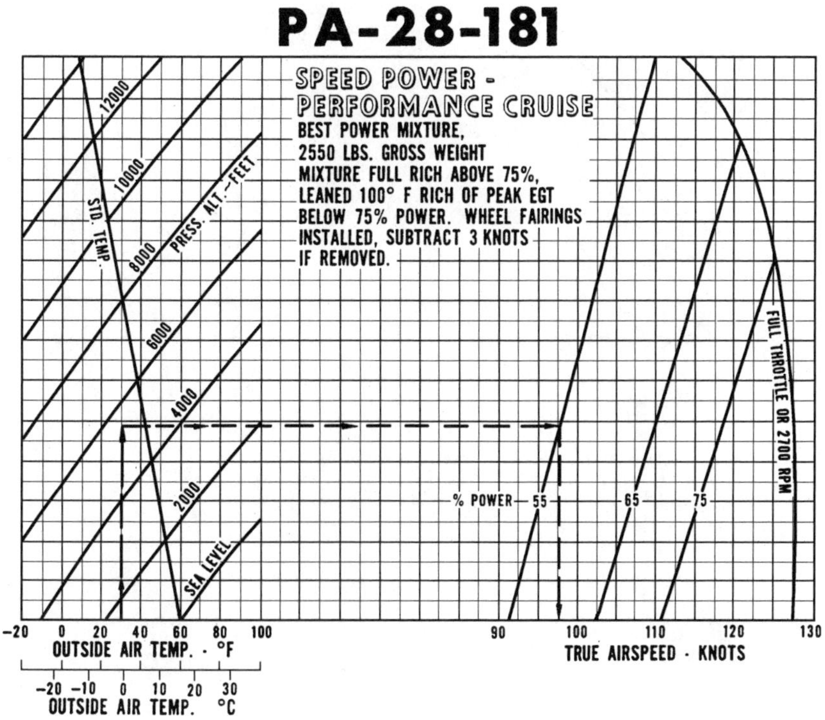

Example:
Cruise pressure altitude: 5500 ft.
Cruise OAT: 30°F
Power: 55%
True airspeed: 97.5 knots

Fig. B-10.

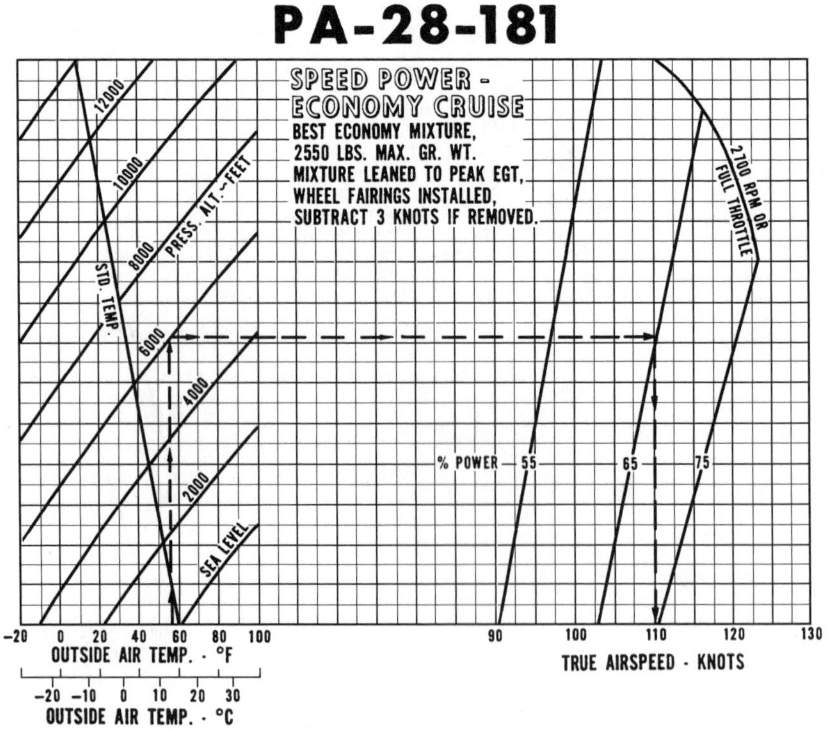

Example:
Cruise pressure altitude: 6000 ft.
Cruise OAT: 55°F
Power: 65%
True airspeed: 110 knots

Fig. B-11.

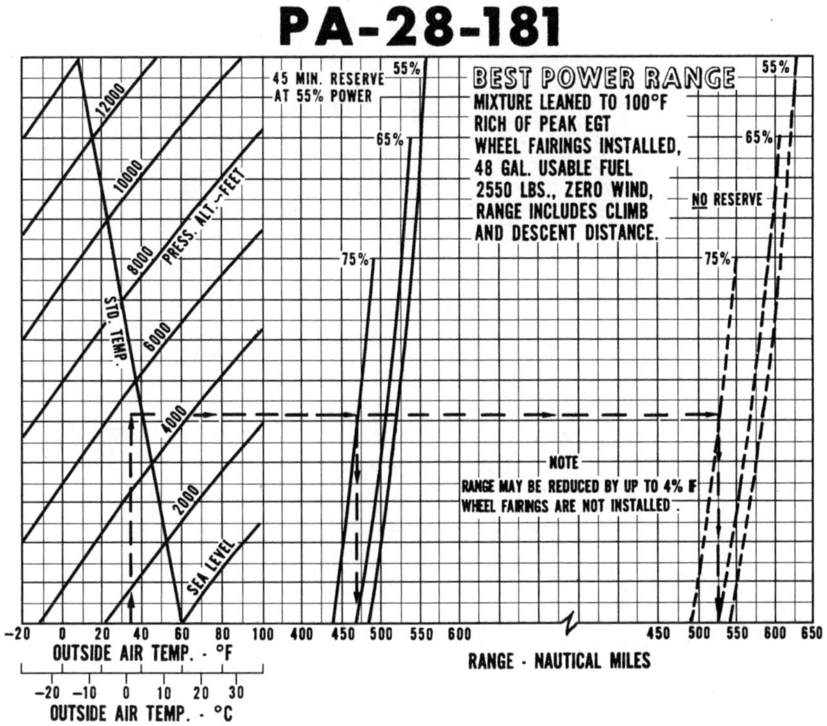

Example:
Cruise pressure altitude: 5500 ft.
Cruise OAT: 35°F
Power setting: 75%
Range (with reserve): 470 nautical miles
Range (no reserve): 525 nautical miles

Fig. B-12.

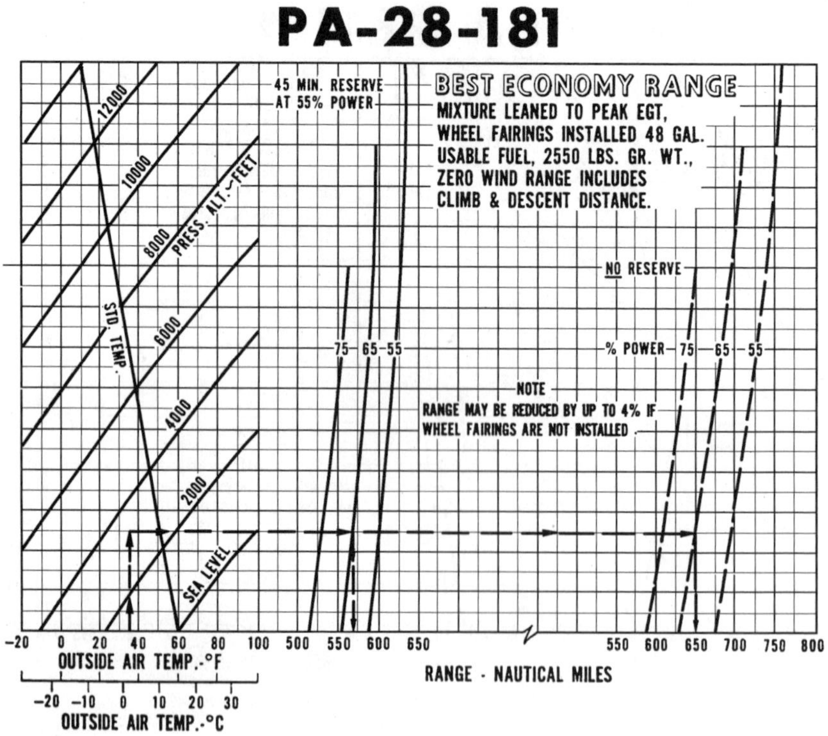

Example:
Cruise pressure altitude: 3000 ft.
Cruise OAT: 35°F
Power setting: 65%
Range (with reserve): 570 nautical miles
Range (no reserve): 650 nautical miles

Fig. B-13.

PA-28-181

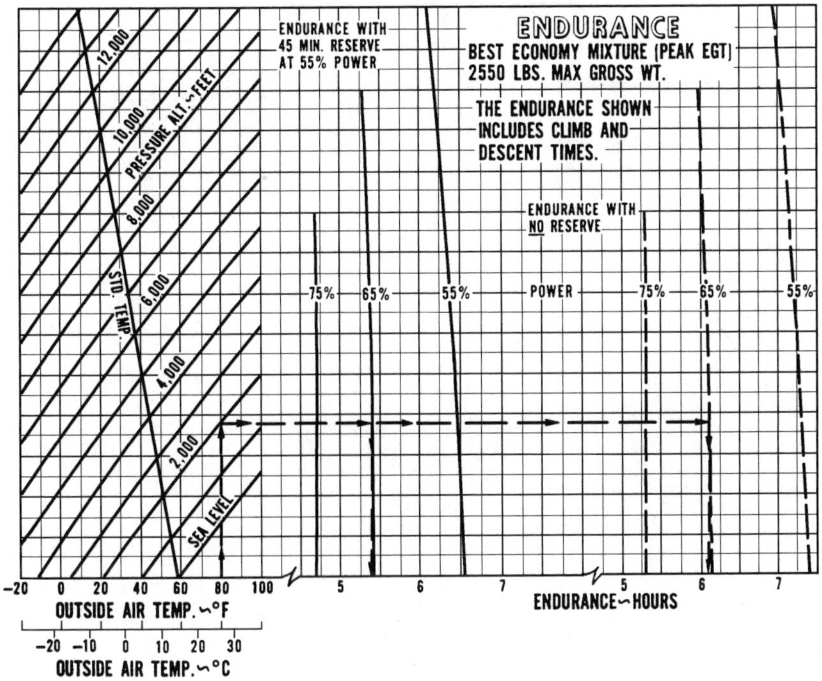

Example:
- Cruise pressure altitude: 2000 ft.
- Cruise OAT: 80°F
- Power setting: 65%
- Endurance (with reserve): 5.37 hrs.
- Endurance (no reserve): 6.1 hrs.

Fig. B-14.

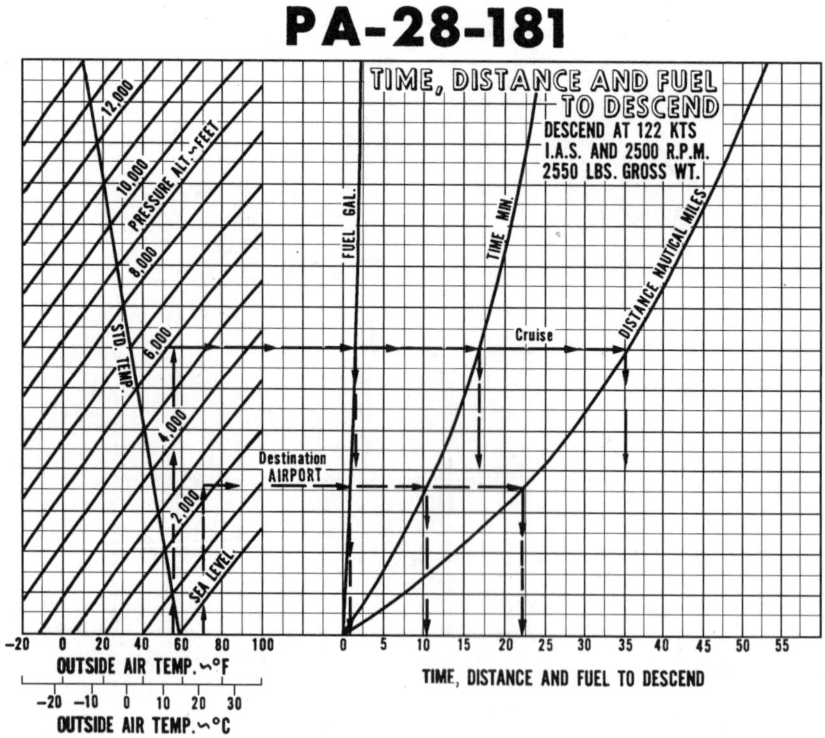

Example:
- Cruise pressure altitude: 6000 ft.
- Cruise OAT: 55°F
- Destination airport pressure altitude: 2300 ft.
- Destination airport temperature: 70°F
- Fuel to descend: (1.7 gal. minus 1 gal.) = .7 gal.
- Time to descend: (17 min. minus 10.5 min.) = 6.5 min.
- Distance to descend (35 miles minus 22 miles) = 13 nautical miles

Fig. B-15.

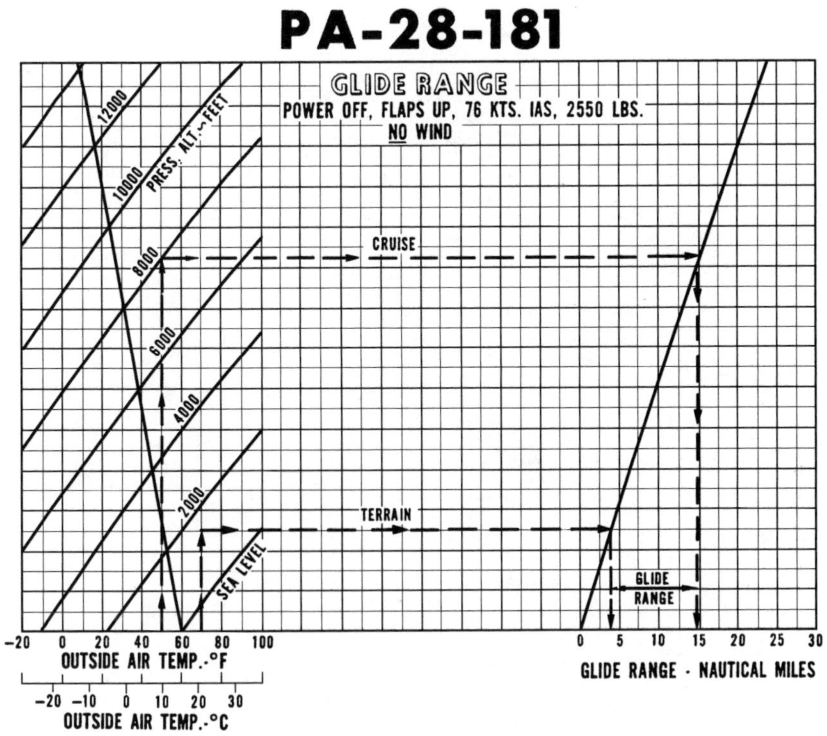

Example:
- Cruise pressure altitude: 8000 ft.
- Cruise OAT: 50°F
- Terrain pressure altitude: 1500 ft.
- Terrain temperature: 70°F
- Glide Range: 14.5 miles minus 3.5 miles = 11 nautical miles

Fig. B-16.

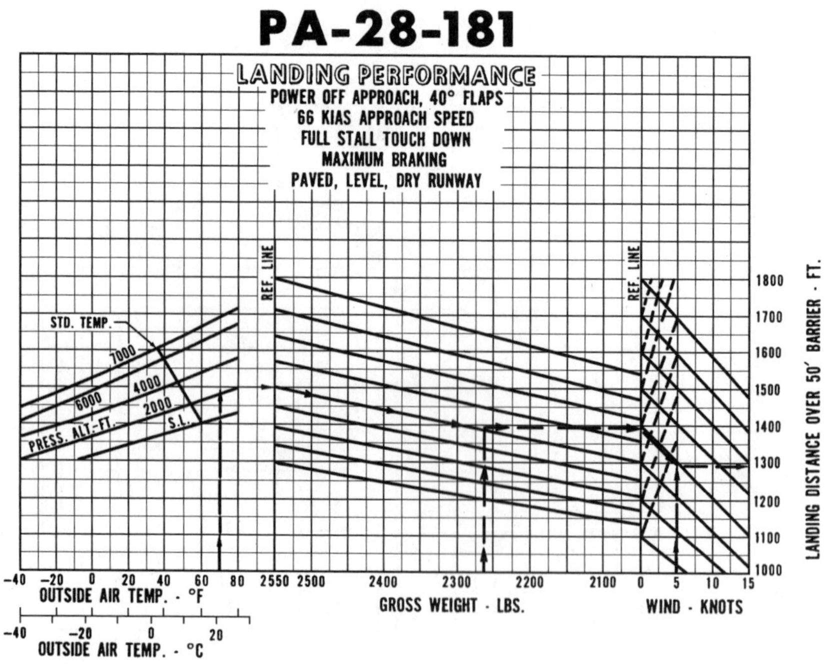

Example:
Airport pressure altitude: 2300 ft.
Gross weight: 2264
Temperature: 70°F
Wind: 5 knots (headwind)
Landing distance: 1290 ft.

Fig. B-17.

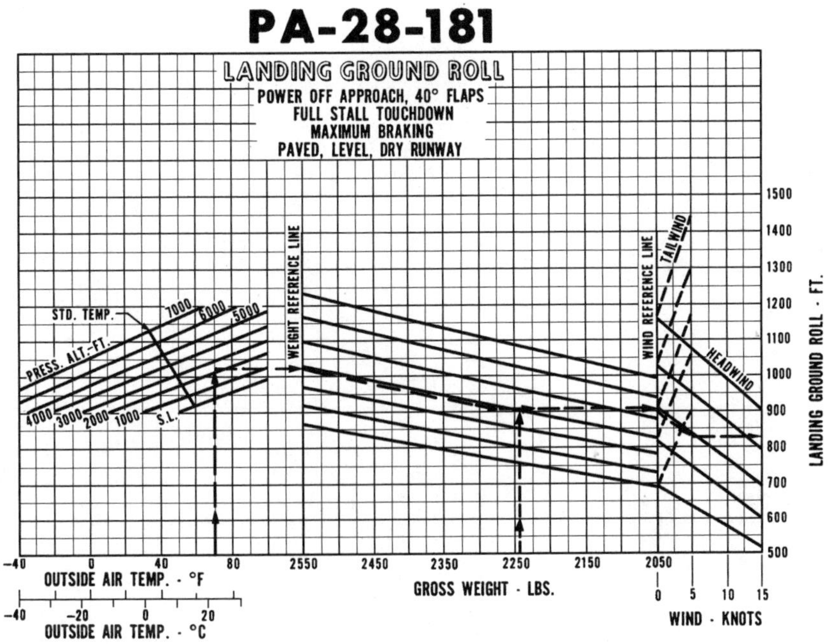

Example:
Airport pressure altitude: 2300 ft.
Airport temperature: 70°F
Gross weight: 2264 lbs.
Wind: 5 knots (headwind)
Ground roll: 825 ft.

Fig. B-18.

CESSNA
MODEL 152

AIRSPEED CALIBRATION

CONDITIONS:
Power required for level flight or maximum rated RPM dive.

FLAPS UP											
KIAS	40	50	60	70	80	90	100	110	120	130	140
KCAS	46	53	60	69	78	88	97	107	117	127	136
FLAPS 10°											
KIAS	40	50	60	70	80	85	---	---	---	---	---
KCAS	44	52	61	70	80	84	---	---	---	---	---
FLAPS 30°											
KIAS	40	50	60	70	80	85	---	---	---	---	---
KCAS	43	51	61	71	82	87	---	---	---	---	---

Fig. B-19.

CESSNA
MODEL 152

STALL SPEEDS

CONDITIONS:
Power Off

NOTES:
1. Altitude loss during a stall recovery may be as much as 160 feet.
2. KIAS values are approximate and are based on airspeed calibration data with power off.

MOST REARWARD CENTER OF GRAVITY

WEIGHT LBS	FLAP DEFLECTION	ANGLE OF BANK 0°		30°		45°		60°	
		KIAS	KCAS	KIAS	KCAS	KIAS	KCAS	KIAS	KCAS
1670	UP	36	46	39	49	43	55	51	65
	10°	36	43	39	46	43	51	51	61
	30°	31	41	33	44	37	49	44	58

MOST FORWARD CENTER OF GRAVITY

WEIGHT LBS	FLAP DEFLECTION	ANGLE OF BANK 0°		30°		45°		60°	
		KIAS	KCAS	KIAS	KCAS	KIAS	KCAS	KIAS	KCAS
1670	UP	40	48	43	52	48	57	57	68
	10°	40	46	43	49	48	55	57	65
	30°	35	43	38	46	42	51	49	61

Fig. B-20.

TAKEOFF DISTANCE

SHORT FIELD

CONDITIONS:
Flaps 10°
Full Throttle Prior to Brake Release
Paved, Level, Dry Runway
Zero Wind

NOTES:
1. Short field technique as specified in Section 4.
2. Prior to takeoff from fields above 3000 feet elevation, the mixture should be leaned to give maximum RPM in a full throttle, static runup.
3. Decrease distances 10% for each 9 knots headwind. For operation with tailwinds up to 10 knots, increase distances by 10% for each 2 knots.
4. For operation on a dry, grass runway, increase distances by 15% of the "ground roll" figure.

WEIGHT LBS	TAKEOFF SPEED KIAS		PRESS ALT FT	0°C		10°C		20°C		30°C		40°C	
	LIFT OFF	AT 50 FT		GRND ROLL	TOTAL TO CLEAR 50 FT OBS	GRND ROLL	TOTAL TO CLEAR 50 FT OBS	GRND ROLL	TOTAL TO CLEAR 50 FT OBS	GRND ROLL	TOTAL TO CLEAR 50 FT OBS	GRND ROLL	TOTAL TO CLEAR 50 FT OBS
1670	50	54	S.L.	640	1190	695	1290	755	1390	810	1495	875	1605
			1000	705	1310	765	1420	825	1530	890	1645	960	1770
			2000	775	1445	840	1565	910	1690	980	1820	1055	1960
			3000	855	1600	925	1730	1000	1870	1080	2020	1165	2185
			4000	940	1775	1020	1920	1100	2080	1190	2250	1285	2440
			5000	1040	1970	1125	2140	1215	2320	1315	2525	1420	2750
			6000	1145	2200	1245	2395	1345	2610	1455	2855	1570	3125
			7000	1270	2470	1375	2705	1490	2960	1615	3255	1745	3590
			8000	1405	2800	1525	3080	1655	3395	1795	3765	1940	4195

Fig. B-21.

CESSNA
MODEL 152

TIME, FUEL, AND DISTANCE TO CLIMB

MAXIMUM RATE OF CLIMB

CONDITIONS:
Flaps Up
Full Throttle
Standard Temperature

NOTES:
1. Add 0.8 of a gallon of fuel for engine start, taxi and takeoff allowance.
2. Mixture leaned above 3000 feet for maximum RPM.
3. Increase time, fuel and distance by 10% for each 10°C above standard temperature.
4. Distances shown are based on zero wind.

WEIGHT LBS	PRESSURE ALTITUDE FT	TEMP °C	CLIMB SPEED KIAS	RATE OF CLIMB FPM	FROM SEA LEVEL		
					TIME MIN	FUEL USED GALLONS	DISTANCE NM
1670	S.L.	15	67	715	0	0	0
	1000	13	66	675	1	0.2	2
	2000	11	66	630	3	0.4	3
	3000	9	65	590	5	0.7	5
	4000	7	65	550	6	0.9	7
	5000	5	64	505	8	1.2	9
	6000	3	63	465	10	1.4	12
	7000	1	63	425	13	1.7	14
	8000	-1	62	380	15	2.0	17
	9000	-3	62	340	18	2.3	21
	10,000	-5	61	300	21	2.6	25
	11,000	-7	61	255	25	3.0	29
	12,000	-9	60	215	29	3.4	34

Fig. B-22.

CESSNA
MODEL 152

CRUISE PERFORMANCE

CONDITIONS:
1670 Pounds
Recommended Lean Mixture (See Section 4, Cruise)

NOTE:
Cruise speeds are shown for an airplane equipped with speed fairings which increase the speeds by approximately two knots.

PRESSURE ALTITUDE FT	RPM	20°C BELOW STANDARD TEMP			STANDARD TEMPERATURE			20°C ABOVE STANDARD TEMP		
		% BHP	KTAS	GPH	% BHP	KTAS	GPH	% BHP	KTAS	GPH
2000	2400	---	---	---	75	101	6.1	70	101	5.7
	2300	71	97	5.7	66	96	5.4	63	95	5.1
	2200	62	92	5.1	59	91	4.8	56	90	4.6
	2100	55	87	4.5	53	86	4.3	51	85	4.2
	2000	49	81	4.1	47	80	3.9	46	79	3.8
4000	2450	---	---	---	75	103	6.1	70	102	5.7
	2400	76	102	6.1	71	101	5.7	67	100	5.4
	2300	67	96	5.4	63	95	5.1	60	95	4.9
	2200	60	91	4.8	56	90	4.6	54	89	4.4
	2100	53	86	4.4	51	85	4.2	49	84	4.0
	2000	48	81	3.9	46	80	3.8	45	78	3.7
6000	2500	---	---	---	75	105	6.1	71	104	5.7
	2400	72	101	5.8	67	100	5.4	64	99	5.2
	2300	64	96	5.2	60	95	4.9	57	94	4.7
	2200	57	90	4.6	54	89	4.4	52	88	4.3
	2100	51	85	4.2	49	84	4.0	48	83	3.9
	2000	46	80	3.8	45	79	3.7	44	77	3.6
8000	2550	---	---	---	75	107	6.1	71	106	5.7
	2500	76	105	6.2	71	104	5.8	67	103	5.4
	2400	68	100	5.5	64	99	5.2	61	98	4.9
	2300	61	95	5.0	58	94	4.7	55	93	4.5
	2200	55	90	4.5	52	89	4.3	51	87	4.2
	2100	49	84	4.1	48	83	3.9	46	82	3.8
10,000	2500	72	105	5.8	68	103	5.5	64	103	5.2
	2400	65	99	5.3	61	98	5.0	58	97	4.8
	2300	58	94	4.7	56	93	4.5	53	92	4.4
	2200	53	89	4.3	51	88	4.2	49	86	4.0
	2100	48	83	4.0	46	82	3.9	45	81	3.8
12,000	2450	65	101	5.3	62	100	5.0	59	99	4.8
	2400	62	99	5.0	59	97	4.8	56	96	4.6
	2300	56	93	4.6	54	92	4.4	52	91	4.3
	2200	51	88	4.2	49	87	4.1	48	85	4.0
	2100	47	82	3.9	45	81	3.8	44	79	3.7

Fig. B-23.

LANDING DISTANCE

SHORT FIELD

CONDITIONS:
Flaps 30°
Power Off
Maximum Braking
Paved, Level, Dry Runway
Zero Wind

NOTES:

1. Short field technique as specified in Section 4.
2. Decrease distances 10% for each 9 knots headwind. For operation with tailwinds up to 10 knots, increase distances by 10% for each 2 knots.
3. For operation on a dry, grass runway, increase distances by 45% of the "ground roll" figure.

WEIGHT LBS	SPEED AT 50 FT KIAS	PRESS ALT FT	0°C		10°C		20°C		30°C		40°C	
			GRND ROLL	TOTAL TO CLEAR 50 FT OBS	GRND ROLL	TOTAL TO CLEAR 50 FT OBS	GRND ROLL	TOTAL TO CLEAR 50 FT OBS	GRND ROLL	TOTAL TO CLEAR 50 FT OBS	GRND ROLL	TOTAL TO CLEAR 50 FT OBS
1670	54	S.L.	450	1160	465	1185	485	1215	500	1240	515	1265
		1000	465	1185	485	1215	500	1240	520	1270	535	1295
		2000	485	1215	500	1240	520	1270	535	1300	555	1330
		3000	500	1240	520	1275	540	1305	560	1335	575	1360
		4000	520	1275	540	1305	560	1335	580	1370	600	1400
		5000	540	1305	560	1335	580	1370	600	1400	620	1435
		6000	560	1340	580	1370	605	1410	625	1440	645	1475
		7000	585	1375	605	1410	625	1440	650	1480	670	1515
		8000	605	1410	630	1450	650	1480	675	1520	695	1555

Fig. B-24.

Bibliography

Anderson, John D., Jr., *Aircraft Performance and Design,* McGraw-Hill, Inc., New York, NY, 1999.

Askue, Vaughan, *Flight Testing Homebuilt Aircraft,* Iowa State University Press, Ames, IA, 1992.

Dole, Charles E., *Flight Theory and Aerodynamics, 2d Edition,* John Wiley & Sons, Inc., New York, NY, 2000.

FAA, *Amateur-Built Aircraft Flight Testing Handbook,* U.S. Government Printing Office, Washington, DC, 1990.

Hamlin, B., *Flight Testing Conventional and Jet-Propelled Airplanes,* The Macmillan Co., New York, NY, 1946.

Herrington, R. M., and P. E. Shoemacher, *Flight Test Engineering Handbook,* AF Technical Report No. 6273, USAF Systems Command, Edwards AFB, CA, 1964.

McCormick, Barnes W., *Aerodynamics, Aeronautics, and Flight Mechanics, 2d Edition,* John Wiley & Sons, Inc., New York, NY, 1995.

Smith, Hubert C., *Introduction to Aircraft Flight Test Engineering,* Jeppesen Sanderson, Englewood, CO, 1981.

Smith, H. C. "Skip," *The Illustrated Guide to Aerodynamics, 2d Edition,* McGraw-Hill, Inc., New York, NY, 1992.

Ward, Donald T., *Introduction to Flight Test Engineering,* Elsevier Science Publishers B.V., Amsterdam, The Netherlands, 1993.

Index

Note: boldface numbers indicate illustrations; italic *t* indicates a table.

abbreviations and acronyms, 27, 181–182
absolute ceiling, 91, **95**, 157
absolute range calculation, 118–119
aerobatic flight, 11
aerodynamics, 62–63, **63**, 173–174
air (*See* atmospheric effects)
airspace for flight tests, 11
airspeed, 11, 33–50, 141–143, **142, 144**
  altitude vs., 43
  angle of attack vs., 35, 36
  calibrated (CAS), 33, 48–49
  climb performance vs., 48, 91, 94, 156
  crosswind and drift vs., 40–41, **41**, 45, 46
  cruise performance vs., 106–107
  density correction for true airspeed in, 46–48
  density, density altitude vs., 175
  descent performance and, 124–125
  drag vs., 175
  dynamic pressure and, 48
  equivalent (EAS), 47
airspeed (*Cont.*):
  errors in, 33–35
  gauge error in, 33–34
  glide performance vs., 165
  Global Positioning System (GPS) method in, 43–46, 49–50, 175
  graphing or plotting results of, 41–42, **42**
  groundspeed and, 36, 37, 39–40, 44, 45
  headwind or tailwind vs., 38–39, **39**
  indicated airspeed (IAS), 13, 33
  indicators for, 13, **14**, 16, 34, 46
  landing performance vs., 128–129, 131–132, 133
  landmark passage in, using wing edge or wheel fairing as marker for, 42–43, **43, 44**
  pacer method in, 36–37
  photogrid method for, 36
  pitot tube in, 34–35, **34, 35**
  position error in, 34–35, 36
  power settings vs., 175–177, **176**
  speed enhancements for, 174–177, **176**
  speed-course method in, 37–43, **38**, 48–49

airspeed (*Cont.*):
stall speed vs., 48, 54, 57, 59
static port in, 34–35, **36**
takeoff performance vs., 61, 70
test procedure for, 36–43
"trailing bomb" method in, 37
true (TAS), 13, 37–38, 44–48
wind vs., 38
airspeed indicators, 13, **14**, 16, 34, 46
altimeters, 5, 14–15, **15**, 16
altitude, 11
airspeed calibration vs., 43
altimeters for, 5, 14–15, **15**, 16
atmospheric pressure vs., 3–5, **5**, 14, 16
climb performance vs., 87–88, **88**, 92–93, 96–100, 153–154, **155**, 157, **158**
cruise performance vs., 102–106, **103, 107**, 108–110, **108**
density of air and, 1–2, 5–10, **6, 7, 8, 9**, 14
descent performance vs., 126, **127**
glide performance vs., 161, 165–167, **167**, 168
landing performance vs., 170
minimum safe altitudes, 11
takeoff performance vs., 72–75, 151–152, **153**
temperature vs., 3
true, 14
anemometers, 70–72, **71**
angle of attack
airspeed calibration vs., 35, 36
stall speed vs., 51–54, **52, 54**
takeoff performance and, 61–62
angle-of-climb speed (V_x), 77–78
atmospheric effects, 1–10
atmospheric pressure vs. temperature, 2
density of air, density altitude, 1–2, 5–10, **6, 7, 8, 9**
density ratio (p/p_0), 7–10, **8**
mass density or slugs, 3
mean sea level (MSL) and, 2
atmospheric effects (*Cont.*):
pressure altitude, 3–5, **5**, 3
standard atmospheric properties, 3, 4*t*
standard density, 2–3
standard lapse rate (temperature vs. altitude), 3
temperature vs. density altitude, 2
weight of air, 1
atmospheric pressure
altitude vs., 3–5, **5**
density altitude vs., **184**
temperature vs., 2

banking, stall speed vs., 54–55, **56**, 59
Bernoulli's theorem, 51
brake horsepower
climb performance and, 87, 88
cruise performance and, 102
braking, and landing performance, 130–132, 168

calibrated airspeed (CAS), 48–49, 142
climb performance vs., 94
descent performance and, 124–126
stall speed vs., 54
takeoff performance vs., 70
ceiling, for climb performance, 91, **95** 157, **159**
climb performance, 83–100
airspeed vs., 48, 91, 94, 156
altitude vs., 87–88, **88**, 92–93, 96–100, 153–154, **155**, 157, **158**
brake horsepower and, 87, 88
ceiling in, service and absolute, 91, **95**, 157, **159**
density, density altitude vs., 93–94, **95**, 98–100, 153–154, **155**,, 157, **158**, 177–178
drag vs., 83–86, **84, 85, 86**
dynamic pressure and, 85
example calculations for, 95–96
forward velocity in, 89–80, **90**
friction and, 87
landing gear vs., up or down, 94–95
lift and, 85, **85**, 89

climb performance (*Cont.*):
lift-to-drag (L/D) ratio in, 89
modifications to airplane vs. for, 177–178
power available for, 87–90
power curve in, 87, **87**
power required in, 86–87, **87**
power settings for, 87
propeller efficiency in, 89, **89**
rate-of-climb speed (V_y) in, 83, 90–91, 93–95, **94, 95**, 99–100, 152–157, **154**, 177–178
shaft horsepower and, 87, 88
temperature vs., 92–93, 98
test procedure for, 91–95
thrust horsepower and, 88–90, **90**
time and distance to climb calculation in, 96–100, **97, 98, 99, 100**
weight vs., 91–92, 156–157
wind vs., 92
computer calculations, 139–172
for airspeed calibration, 141–143, **142, 144**
for cruise performance, 157–161, **160**
Denalt Performance Computer in, 74–75, **74**
for density, altitude density, 140
differentiation in, 147
E6B type computer in, 6–9, **9**, 25, **25**
for glide performance, 161–168, **164**
graphing results of, 140–141
for landing performance, 168–171, **172**
Microsoft Excel in, 139
for rate-of-climb speed, 152–157, **154**
for stall speed, 143–145, **145, 146**
for takeoff performance, 145–152, **147, 148, 150**
TechStar computer in, 25
continuity principle in airfoil design, 51
convective currents or thermals, 21–24, **23**
conversion factors, 183
crosswind and drift vs. airspeed calibration, 40–41, **41**, 45, 46
cruise performance, 101–120, 157–161, **160**
absolute range calculation in, 118–119
airspeed vs., 106–107
altitude vs., 102–110, **103, 107, 108**
brake horsepower in, 102
cruise speed in, 101
density, density altitude vs., 102, 106, 108–109, **119**, 160–161, **162**
drag vs., 112–113
economy cruise power setting in, 102
endurance calculation in, 119
example calculations for, 120
exhaust gas temperature (EGT) gauge in, 105, 114, **114**
fuel consumption vs. range in, 111, 113–118
fuel-air mixture control and, 105
horsepower ratings in, 102
maximum speed test procedure for, 107–110, **108, 109, 110**
modifications to airplane vs., 173
normal cruise power setting in, 102
performance cruise power setting in, 102
power determination in, 101–105, **103**, 104*t*
power-required curves in, 111–112, **112**
range calculation in, 101, 110–118, **112**, 160–161, **163**
range test procedures in, 113–118
speed enhancements, 174–177, **176**
takeoff power rating and, 108–109
temperature vs., 106–107
test procedure for, 105–107
weight vs. range in, 112–113, 157–161
curves, plotting curves on graphs, 30

data gathering methods, 12–18
data sheets, 13, **13**, 185
deceleration chutes, 130
delta, as increment sign, 27

Denalt Performance Computer, 74–75, **74**
density, density altitude, 1–2, 5–10, **6, 7, 8, 9**, 7–10, **8**, 27, 140
  airspeed vs., 175
  climb performance vs., 93–94, **95**, 98–100, 153–154, **155**, 157, **158**, 177–178
  cruise performance vs., 102, 106, 108–109, **119**, 160–161, **162**
  glide performance vs., 161, 165, **167**
  landing performance vs., 133–134, 170
  stall speed vs., 53
  takeoff performance vs., 73–75, **76**
  temperature and pressure vs., **184**
  true airspeed vs., 46–48
descent performance (*See also* landing), 121–138
  airspeed and, 124–125
  altitude vs., 126, **127**
  drag vs., 126
  example calculations for, 135–136
  glide distance in, 122–123, **123**
  glide ratio in, 123–127, **125, 127**
  landing performance in, 127–135
  lift-to-drag (L/D) ratio in, 123–124, 124–127, **125, 127**
  power available in, 121
  power-off glide in, power decrement in, 120–121, **121**
  rate-of-sink in, 121–122
  thrust horsepower in, 121, 122–123
  weight vs. glide in, 126
  winds vs., 126–127
differentiation in calculations, 147
drag, 174
  airspeed vs., 175
  climb performance vs., 83–86, **84, 85, 86**
  cruise performance vs., 112–113
  descent performance vs., 126
  landing performance vs., 130
  stall speed vs., 180
  takeoff performance vs., 62
drift vs. airspeed calibration, 40–41, **41**
dynamic pressure, 48, 85

E6B type computer, 6–9, **9**, 25, **25**
economy cruise power setting, 102
emergency procedures, 12
empty weight, 19
endurance calculation, 119
equations used in flight testing, 25–27
equivalent airspeed (EAS), 47
evaluating modification effects, 173–180
  climb performance and, 177–178
  speed enhancements, 174–177, **176**
  stall speed, 178–180
Excel, 139
exhaust gas temperature (EGT) gauge, cruise performance, 105, 114, **114**

Federal Aviation Regulations (FARs) and tests, 11
flare distance, 128, **129**, 129–130
form drag, 84, **84**
friction
  climb performance vs., 87
  landing performance vs., 130
  takeoff performance vs., 62, 63
fuel consumption, 15–16
  fuel flow gauge in, 115, **115**
  fuel flow meters in, 15–16, **17**
  fuel pump meter in, 115, **116**
  gross weight determination for, 18–21, 20*t*
  range and cruise performance vs., 111, 113–118
fuel-air mixture control, cruise performance and, 105

gap seals, 173, **175**
gauge error in airspeed, 33–34
general flight test procedures, 11–31
  for airspeed calibration, 13
  convective currents, thermals, and turbulence vs., 21–24, **23**
  data gathering methods for, 12–18
  data sheets for, 13, **13**, 185
  density, density altitude and, 14
  equations used in, 25–27
  fuel consumption measurement for, 15–16

general flight test procedures (*Cont.*):
graphing or presentation of data from, 27–31
gross weight determination for, 18–21
manifold pressure gauge in, 16, 17, **17**
power settings in, 16
pre-test preparations and planning for, 11–12, **12**
pressure altitude for, 14, 16
recording results of, 17
single-place airplanes, special considerations for, 17–18
tabular presentation of data in, 30–31, 31*t*
tachometers for, 16, 17, **17**
temperature measurement for, 15
time of day for, 22–23
weather minimums for, 11, 21–24
winds vs., 24–25
glide distance, 165
glide performance, 161–163, **164**
airspeed vs., 165
altitude vs., 161, 165–168, **167**
density, density altitude vs., 161, 165, **167**
glide distance in, 165
glide speed in, 168
lift-to-drag (L/D) ratio in, 165, **166**
rate-of-climb speed in, 161
temperature vs., 161
weight vs., 161, 165–168
glide ratio in, 123–127, **125, 127**
glide speed, 168
Global Positioning System (GPS)
ground track of, 44–45
method in airspeed calibration, 43–46, 49–50, 175
graphing or presentation of data, 27–31, 140–141
gravity, 1, 62–63
gross weight determination (*See also* weight), 18–21
ground roll, 130–131, 133–135, 170
groundspeed, 36, 37, 39–40, 44, 45, 142
haze, 23
headwind or tailwind
airspeed calibration vs., 38–39, **39**
takeoff performance vs., 72, **73**
hinge fairings, 173, **175**
horsepower, 87, 102

indicated airspeed (IAS), 13, 142
climb performance vs., 94
descent performance vs., 124–125
landing performance vs., 132
induced drag, 85, **85**, 112
induced power, 86–87
inertia
landing performance vs., 130
takeoff performance vs., 62

knots vs. miles per hour, 140, 183

landing performance, 127–135, 168–171, **172**
airspeed vs., 128–129, 131–133
altitude vs., 170
braking in, 130–132, 168
density, density altitude vs., 133–134, 170
drag in, 130
example calculations for, 136–137
flare distance required for, 128, **129**, 129–130
forces acting on, 130–131, **131**
friction and, 130
ground roll in, 130–131, 133–135, 170
inertia and, 130
lift-to-drag (L/D) ratio in, 129–130, 135, **136**, 170–171
runway distance required for, 127–128, **129**, 133, 169–170, **172**
stall speed and, 128–130, 135, 168–169, 178, 180
stopping point for, 132, 132,168
test procedures for, 131–135
touchdown point for, 132, **132**, 168
weight vs., 128, 133–134, **134**, 168–169
winds vs., 133, 168–170

leaning (*See also* fuel-air mixture control), 105
lift
  climb performance vs., 85, **85**, 89
  stall speed vs., 51–54, **52, 54**, 55
lift-to-drag (L/D) ratio
  climb performance vs., 89
  descent performance vs., 123–127, **125, 127**
  glide performance vs., 165, **166**
  landing performance vs., 129–130, 135, **136**, 170–171
liftoff point calculation for takeoff performance, 64
line graphs, 27–28, **28**, 28*t*, **29, 30, 31**

manifold pressure gauge, 16, 17, **17**
mass density or slugs, 3
maximum cruise speed test, 107–110, **108, 109, 110**
mean sea level (MSL), 2
metric conversion factors, 183
Microsoft Excel, 139
miles per hour vs. knots, 140, 183
modifications (*See* evaluating modification effects)
mountain waves, 23–24, **24**

nautical miles vs. statute miles, 165, 183
night flight tests, 22
normal cruise power setting, 102

obstruction clearance in takeoff performance, 75–78, **77**
obstructions, turbulence caused by, 23–24
outside air temperature (OAT) gauge, 6, 15, **16**, 16

pacer method in airspeed calibration, 36–37
parasite drag, 84, 85
parasite power, 86
performance cruise power setting, 102
performance data, 31
photo recording method for takeoff performance, 66–68, **67**
photogrid method for airspeed calibration, 36
pitot tube in airspeed calibration, 34–35, **34, 35**
position error in airspeed calibration, 34–36
power available
  in climb performance, 87–90
  in descent performance, 121
power curve for climb performance, 87, **87**
power settings, 16
  airspeed vs., 175–177, **176**
  climb performance, 87
  cruise performance, 101–105, **103**, 104*t*
power required
  in cruise performance, 111–112, **112**
  in climb performance, 86–87, **87**
powers of numbers, 26
pressure altitude, 3–5, **5**, 14, 16
pressure drag, 84
propeller efficiency, climb performance and, 89, **89**
propellers, 16

range calculation, 101, 110–118, **112**, 160–161, **163**
rate-of-climb speed (V_y), 77–78, 83, 90–91, 93–95, **94, 95**, 99–100, 152–157, **154**, 177–178
  in glide performance, 161
  modifications to airplane vs., 173
rate-of-climb tests, 22, 23, 29, **29**
rate-of-descent tests, 22
rate-of-sink, in descent performance, 121–122
recording results, 17
retarding forces, in takeoff performance, 62
runway distance required for landing, 127–128, **129**, 133, 169–170, **172**
runway distance required for takeoff, 61, 147–149, **147, 148**

runway distance vs. time, 27–28, **28**, 28*t*
runway gradient vs. takeoff performance, 62–63, 75
runway surface condition vs. takeoff performance, 63

sailplanes, 22
seatbelts and restraints, 12
service ceiling, 91, **95**
shaft horsepower, climb performance and, 87, 88
single-place airplanes, special considerations for, 17–18
skin friction drag, 84, **84**
slug, 3
speed enhancements, 174–177, **176**
speed-course method in airspeed calibration, 37–43, **38**, 48–49
spotters for flight tests, 11
square roots of numbers, 26
stall speed (V_S), 11, 26, 30, **31**, 31*t*, 51–59, 143–145, **145, 146**
  airspeed vs., 48, 54, 57, 59
  angle of attack and, 51–54, **52, 54**
  bank effects on, 54–55, **56**, 59
  Bernoulli's theorem and, 51
  density, density altitude vs., 53
  drag vs., 180
  entering the stall to determine, 55–57, **57**
  example calculations for, 59
  gross weight determination for, 18–21
  landing performance vs., 128–130, 135, 168–169, 178, 180
  lift vs., 51–54, **52, 54**, 55
  modifications to airplane vs., 173, 178–180
  takeoff performance vs., 61, 70, 149, 178, 180
  test procedures for, 55–59
  velocity vs., 53
  vortex generators and, 178–179, **179**
  weight vs., 53–54, 57, **58**, 143–145, **145, 146**, 179–180
standard density, 2–3
standard lapse rate (temperature vs. altitude), 3
standard weight (W_S), 18–21, 27
static port, in airspeed calibration, 34–35, **36**
statute miles vs. nautical miles, 165, 183
stopping point, in landing performance, 168
subscripts, 26
symbols, 181–182

tachometers, 16, 17, **17**
takeoff performance, 61–81, 145–152, **147, 148, 150**
  airspeed vs., 61, 70
  altitude correction in, 72–75, 151–152, **153**
  angle of attack and, 61, 62
  angle-of-climb speed (V_x), 77–78
  density, density altitude vs., 73–75, **76**
  drag vs., 62
  example calculations for, 78–81, 79*t*, **79**
  forces acting on airplane in, 62–63, **63**
  frictional force vs., 62–63
  gravity vs., 62–63
  inertia vs., 62
  liftoff point calculation for, 64
  obstruction clearance and, 75–78, **77**
  photo recording method for, 66–68, **67**
  rate-of-climb speed (V_y), 77–78
  retarding forces vs., 62
  runway distance required for, 61, 147–149, **147, 148**
  runway gradient (slope) vs., 62–63, 75
  stall speed vs., 61, 70, 149, 178, 180
  surface conditions vs., 63
  takeoff speed and, 61
  test procedure for, 63–66, **65**
  thrust and, 62

takeoff performance (*Cont.*):
  velocity calculation in, 68–70, **69, 71**, 149, **150**, 149
  weight vs., 62, 64, 75, **76**, 151–152
  wind correction in, 70–72, **71, 73**, 151
takeoff power rating, cruise performance and, 108–109
takeoff speed, 61
TechStar computer, 25
temperature, 15, 183
  altitude vs., 3
  atmospheric pressure vs., 2
  climb performance vs., 92–93, 98
  convective currents (thermals) and, 21–24, **23**
  cruise performance vs., 106–107
  density altitude vs., 2, **184**
  glide performance vs., 161
  outside air temperature (OAT) gauge for, 6, 15, **16**, 16
thermals (*See* convective currents)
thrust, 62, 174, 175
thrust horsepower
  climb performance vs., 88–90, **90**, 89
  descent performance vs., 121–123
thrust reversers, 130
touchdown point, in landing performance, 168
"trailing bomb" method in airspeed calibration, 37
true airspeed (TAS, V_t), 13, 27, 37–38, 44–48
  climb performance vs., 94, 156
  cruise performance vs., 106–107
  descent performance and, 125
  landing performance and, 132
true airspeed (*Cont.*):
  stall speed vs., **54**
  takeoff performance vs., 70
true altitude,14
turbulence,21–24, **23**, 92, 106

velocity calculation, 53, 68–70, **69, 71**, 149, **150**
VFR conditions for flight tests, 11, 21
viscous drag, 84
vortex generators, 178–179, **179**

warning horns, 12
weather minimums for flight testing, 11, 21–24
weight
  climb performance vs., 91–92, 156–161
  descent performance vs., 126
  fuel consumption vs., 18–21, 20*t*
  glide performance vs., 161, 165–168
  landing performance vs., 128, 133–134, **134**, 168–170
  range vs., 112–113
  stall speed vs., 53–54, 57, **58**, 143–145, **145, 146**, 179–180
  standard, 18–21
  takeoff performance vs., 62, 64, 75, **76**, 151–152
winds, 24
  airspeed calibration vs., 38
  climb performance vs., 92
  descent performance vs., 126–127
  landing performance vs., 133, 168–169
  takeoff performance vs., 70–72, **71, 73**, 151

About the Author

Hubert C. "Skip" Smith is Associate Professor Emeritus of Aerospace Engineering at The Pennsylvania State University, where he has taught aircraft design, performance and aerodynamics for 30 years. He holds B.S. and M.S. degrees in aeronautical engineering from Penn State, and a Ph.D. from the University of Virginia. An instrument-rated commercial pilot and flight instructor, he has been actively flying for over 45 years. He is an Associate Fellow of the American Institute of Aeronautics and Astronautics, has won a number of national awards for outstanding educational achievements, and is the author of another McGraw-Hill book, *The Illustrated Guide to Aerodynamics.*

DISK WARRANTY

This software is protected by both United States copyright law and international copyright treaty provision. You must treat this software just like a book, except that you may copy it into a computer in order to be used and you may make archival copies of the software for the sole purpose of backing up our software and protecting your investment from loss.

By saying "just like a book," McGraw-Hill means, for example, that this software may be used by any number of people and may be freely moved from one computer location to another, so long as there is no possibility of its being used at one location or on one computer while it also is being used at another. Just as a book cannot be read by two different people in two different places at the same time, neither can the software be used by two different people in two different places at the same time (unless, of course, McGraw-Hill's copyright is being violated).

LIMITED WARRANTY

McGraw-Hill takes great care to provide you with top-quality software, thoroughly checked to prevent virus infections. McGraw-Hill warrants the physical diskette(s) contained herein to be free of defects in materials and workmanship for a period of sixty days from the purchase date. If McGraw-Hill receives written notification within the warranty period of defects in materials or workmanship, and such notification is determined by McGraw-Hill to be correct, McGraw-Hill will replace the defective diskette(s). Send requests to:

McGraw-Hill
Customer Services
P.O. Box 545
Blacklick, OH 43004-0545

The entire and exclusive liability and remedy for breach of this Limited Warranty shall be limited to replacement of defective diskette(s) and shall not include or extend to any claim for or right to cover any other damages, including but not limited to, loss of profit, data, or use of the software, or special, incidental, or consequential damages or other similar claims, even if McGraw-Hill has been specifically advised of the possibility of such damages. In no event will McGraw-Hill's liability for any damages to you or any other person ever exceed the lower of suggested list price or actual price paid for the license to use the software, regardless of any form of the claim.

McGRAW-HILL SPECIFICALLY DISCLAIMS ALL OTHER WARRANTIES, EXPRESS OR IMPLIED, INCLUDING, BUT NOT LIMITED TO, ANY IMPLIED WARRANTY OF MERCHANTABILITY OR FITNESS FOR A PARTICULAR PURPOSE.

Specifically, McGraw-Hill makes no representation or warranty that the software is fit for any particular purpose and any implied warranty of merchantability is limited to the sixty-day duration of the Limited Warranty covering the physical diskette(s) only (and not the software) and is otherwise expressly and specifically disclaimed.

This limited warranty gives you specific legal rights; you may have others which may vary from state to state. Some states do not allow the exclusion of incidental or consequential damages, or the limitation on how long an implied warranty lasts, so some of the above may not apply to you.